Fraction Speed Bumps

Fraction Challenges that Enhance and Review Basic Skills

Grades 5-8

About the author

Nate Cattell graduated from Pennsylvania State University with a BA in graphic design. He later earned his Pennsylvania Elementary Teaching Certification and has taught sixth grade in the State College Area School District for 31 years. In 1991, Nate was the Pennsylvania recipient of a Presidential Award for Excellence in Mathematics Teaching. He delights in creating new challenge activities for his students. For Interact, Nate also authored *Challenge Math Projects* (2000), *Geometry Challenge* (2002), and *Algebra Mystery Maze* (2003).

Revised 1/2010

10200 Jefferson Blvd • P.O. Box 802 Culver City, CA 90232
Phone: (800) 359-0961 • www.teachinteract.com
ISBN# 978-1-56004-386-7

Welcome to *Fraction Speed Bumps!*

Fraction Speed Bumps is a unit that allows teams to review fraction theory through exploration, cooperation, and competition. Teams of three will use the numbers provided each day to meet a variety of fraction challenges and earn points. The points may be used during the hands-on competitions. Marble rolling events, a paper racetrack, and a slow marble race are all part of the excitement. The unit has a teacher's guide, which allows the instructor to match the skill level of the class to a fraction challenge page. The Student Guide reviews eighteen fraction concepts in an easy to understand format. The thirteen day unit may be easily extended or shortened.

Standards

The nationwide movement for high standards has not only determined what students should learn, but also mandated that students demonstrate what they know. *Fraction Speed Bumps* is a standards-based program addressing National Math Standards and provides many opportunities for performance assessments. Students apply their math skills and creativity to solve a variety of fraction challenges. The cooperation, peer teaching, and group decision-making address Applied Learning Standards.

National Standards for School Mathematics (NCTM)

Grades 5–8

Standard 1—Problem Solving

- Use problem-solving approaches to investigate and understand mathematical content
- Develop and apply a variety of strategies to solve problems, with emphasis on multistep and non-routine problems
- Verify and interpret results with respect to the original problem situation
- Generalize solutions and strategies to new problem situations
- Acquire confidence in using mathematics meaningfully

Standard 2—Communication

- Reflect on and clarify their thinking about mathematical ideas
- Discuss mathematical ideas and make conjectures and convincing arguments

Standard 3—Reasoning

- Validate their thinking

Standard 5—Number Relationships

- Develop number sense for whole numbers, fractions, and decimals
- Investigate relationships between fractions and decimals

Standard 6—Number Systems and Number Theory

- Extend their understanding of whole number operations to fractions and decimals
- Understand how the basic number operations are related

Standard 7—Computation and Estimation

- Compute with whole numbers, fractions and decimals
- Develop, analyze, and explain procedures for computation and techniques for estimation
- Use estimation to check the reasonableness of results

Table of Contents

Table of Contents

Table of Contents

Masters

Table of Contents

Purpose

Many students work fraction problems without any real fraction understanding. *Fraction Speed Bumps* offers a different approach to reviewing and manipulating fractions. Students, working in teams of three, experiment with fractions, share their findings, and use logical reasoning. They compete against both the fraction challenges and other teams. A wide variety of activities gives the teacher total control over the length of time spent each day, the number of days the unit will last, and finally, which team building competitions to choose. The Teacher Guide provides daily instructions and procedures. The Student Guide provides information that reviews and reinforces many fraction concepts. You may supplement the challenges provided as dictated by the needs of your students. You may also substitute your own choice of fraction problems for some or all of the competitions.

Every activity in *Fraction Speed Bumps* encourages students to use higher level thinking skills. Knowledge, application, analysis, evaluation, and synthesis are all experienced as the students work through the challenges in this unit. By working through *Fraction Speed Bumps*, students will understand and experience the following:

Knowledge

1. Understand relationships involving fractions
2. Solve progressively more difficult problems
3. Devise creative solutions to fraction challenges
4. Understand more clearly how the functions of multiplication and division relate to fractions
5. Use addition, subtraction, multiplication, and division to solve problems
6. Gain a more complete understanding of fraction practices
7. Use metric measurements

Skills

1. Tackle a problem creatively
2. Demonstrate risk-taking related to fractions and design, based on knowledge, estimation, math concepts, and visualization
3. Design competition models
4. Build on previous learning experiences to solve new fraction challenges

Attitudes

1. Confidence to use his/her own approach to create a solution
2. Appreciate the value of collaborative working relationships
3. Delight in seeing how others have solved a problem
4. Increase knowledge and confidence as they share insights and methods with others
5. Strive for a goal
6. Work cooperatively with other students to develop competition strategies
7. Enjoy increased confidence by developing different approaches to solving fractions

Overview

Fraction Speed Bumps is a math unit that allows students to review fraction theory and enhances understanding of fraction relationships through exploration and competition. This competitive series of math challenges enables teams of three to earn points each day. Each event involves planning, logical thinking, cooperation, and a little luck. During the unit, teams amass points and use those points to purchase materials for building speed bumps. During the culmination, all remaining points are used in a design challenge competition. There are many ways for a team to meet with success.

The team competition allows individuals to attempt a different kind of fraction challenge, rather than just working on regular fraction problems. Teams generate answers based on a different challenge each day. These challenges help them see patterns and cement understanding. Students engage in competition that is both challenging and non-threatening. As the competition progresses, students use strategies based on their evaluation of their previous challenge experiences.

The competition assumes that students have worked previously with fractions. Brief fraction review lessons should be included at the beginning of a class period, followed by the team competition. Challenge fraction pages may be assigned as class work or homework, if needed.

There are different suggestions throughout the unit to help you modify the challenge to match the skill level of each class.

The Student Guide is a guide for reviewing and thinking about fractions. Examples are included to enhance and reinforce prior fraction knowledge.

The speed bump events and the culmination are "design challenges" which are hands-on activities. Building materials (paper strips) are purchased with the points earned earlier during the daily fraction competitions. These creative competitive designs are used during classroom team competitions, with the Slow Marble Race as the culminating activity.

Differentiation

This unit offers opportunities for all students through:

1. Visual/spatial: drawing and designing competition model
2. Bodily/kinesthetic: hands-on creations/projects (models)
3. Interpersonal: interacting and planning with others
4. Intrapersonal: setting own goals, independent study, introspection, and use of creative materials

5. Linguistic: using words effectively in evaluations of strategies
6. Logical/mathematical: reasoning, calculating, thinking conceptually and abstractly, and seeing and exploring patterns and number relationships

Special Needs Students

Like all Interact units, *Fraction Speed Bumps* provides differentiated instruction through its various learning opportunities. Students learn and experience the knowledge, skills, and attitudes through all domains of language (reading, writing, speaking, and listening) and math (counting, tallying, computing, etc.). Adjust the level of difficulty to best fit the needs of your students. Allow students to use calculators. Assist special needs students in the activities to utilize their strengths and to succeed. Work together with the Resource Specialist teacher, Gifted and Talented teacher, or other specialist to coordinate instruction.

About the Unit

Fraction Speed Bumps is a departure from traditional fraction problems. Each activity can be part of the long-term team competition or may stand alone as a challenging review lesson. Teams of three use the values provided each day to meet a variety of challenges. There are opportunities for teams to use logical thinking and strategic planning during each competition. The Student Guide includes basic review information for all students. Teams generate specific answers and compete against each other, but all teams earn points for each challenge.

The daily speed bump fraction challenge encourages every team member to work for the best solution. These challenges offer you flexibility. The competition can stop at any problem, or be split over two days. Discussion questions help students reflect on the main concept for the day. You may use the questions for individual written reflections or class discussion. The fraction challenges are varied and offer complicating twists to maintain interest and encourage competition.

This review unit helps students feel more confident and think about fractions differently while seeing the fraction/decimal relationship in a new way.

Fraction speed bump challenges

Fraction speed bump challenges are pages that review basic fraction processes and create a competitive atmosphere. The teacher may modify the pages to meet the needs of each class. Teams should take time to develop strategies for attacking each page.

Fraction speed bumps: Marble rolling

The actual speed bump marble-rolling activities allow teams to use skill to earn additional points. Some only require the rolling of marbles, while others include team discussion, cooperation, and the construction of a speed bump sheet. Select the types of speed bump activities that will work with your class and your schedule.

Slow Marble Race

The Slow Marble Race is the culmination of the unit. The teams use points they have earned to purchase thick paper for building their racetrack. This activity allows for design creativity and thinking "outside the box." A team with a great idea may win even though it has less track length than other teams. Ideas gained from earlier speed bump marble-rolling competitions may be the key to slowing the marble on its descent from three inches. The unit allows you to choose the competition(s) that best fit(s) your teaching schedule for fractions.

Friendly competition

The competition in *Fraction Speed Bumps* is supposed to be friendly and foster camaraderie. As teammates discover the easiest way to solve a challenge, they should be encouraged to share their insights with each other. They are competing not only against the speed bump challenges, but against another teams' strategy.

1. Depending on the team makeup, the point totals may become lopsided. You can make the competition a good experience for all teams by choosing the best option for running a competitive culmination (see page 61).
2. You can use Extension and Homework activities to give some teams the chance to gain extra points, which makes the final event more competitive. Everyone likes to compete against worthy rivals.
3. Students learn that helping teammates and sharing the workload make the team stronger
4. By the final event, most teams will agree that their point total was a combination of math understanding, physical skill, and in some cases…luck

Identifying opponents

Each day teams need to sit in groups of three and be near their opponents so that the symbols and values written on a team sheet may be checked for accuracy.

There are several methods for identifying opponents:

1. Put team names in a hat and pull each out to determine opponents
2. Have the first place team (most points earned) compete against second place, third against fourth, etc. 1↔2 3↔4 5↔6 7↔8
3. You may have an odd number of teams. If so, give each team a number and have one check two, two check three, three check four and the last team will check the first. 1→2→3→4→5→6→7 (7→1)

All teams should get to compete against all other teams during the unit.

Daily Tasks

Some teacher and student tasks are repeated throughout the unit. Since teams will quickly become familiar with the structure of the competition, these tasks are not always included in every set of instructions.

Teacher tasks:

1. Get out student folders
2. Arrange desks in threes
3. Place papers face down before starting
4. Write fraction examples on the board
5. Lead class review discussion as needed before the competition starts
6. Write number on team papers indicating order of finishing
7. Write answer on board to aid students when checking opponent's work
8. Determine, when a team finishes first, if they may work on homework, an Extension sheet, or cut out speed bumps, catching trays, or racetrack segments
9. Collect team folders each day
10. Decide if an extension, homework, or assessment sheet will be assigned for the day

Student tasks:

1. Write team name on papers
2. Write answers neatly on team the speed bump challenge page
3. Make sure that Checkers check opponent's answers and score sheet
4. Understand that a mistake results in a two-point penalty
5. Remind Recorders to write points earned and spent on the score sheet during the competition and at the end of each challenge

Setup Directions

Before You Begin

- Read the entire Teacher Guide and Student Guide. Decide how you will use *Fraction Speed Bumps* in your math curriculum. Modify the procedures outlined for *Fraction Speed Bumps* to suit your teaching preferences and students' needs.
- Study the fraction speed bumps sheets
- Familiarize yourself with the speed bump design competitions
- Use the daily directions to lead your students through the steps for successful friendly competition through the fraction speed bump sheets and speed bump design activities
- Determine how you will use the Slow Marble Race
- Select Extension Activities to supplement your regular curriculum or to keep the competition balanced

Using the Student Guide

The Student Guide introduces students to *Fraction Speed Bumps*. It also includes informational pages of helpful hints for working with fractions. Use the Student Guide to enhance students understanding of fraction concepts. Encourage the use of its information pages during competition.

Planning Your Schedule

- The daily lesson plans recommend 13 days of lessons
- Adjust the timeline to accommodate your own teaching objectives and the capabilities of your students
- If necessary, you can delay using any fraction sheet until your students have a grasp of that fraction concept
- See Adapting *Fraction Speed Bumps* (p. 22) for more suggestions related to modifying your schedule

Grouping Students

Arrange your class into heterogeneous groups of three students. Each day they work as a team, in three roles: Recorder, Setter, and Checker.

Each role has a primary responsibility for the day, but not all roles are required every day. The team must function well to solve the challenge of the day. Roles rotate (change) every day.

Refer to the Cooperative Group Work Rubric when needed, to reinforce good behavior and to give concrete feedback to students who have not yet built a strong team.

Preparing Your Classroom

- You may designate an area on a bulletin board to record team scores
- Students need an area large enough to work together, hold discussions, and construct speed bumps and the final project. Arrange three student desks side by side to create a team "work area." The Recorder sits in the middle desk each day. Since roles rotate, a new person sits in the middle seat each day.
- At the end of each day, insist that teams collect all materials and store them safely in the room. If any materials are too bulky for the team folders, have students put all work in large paper/plastic bags marked with team names.
- Consider making a large poster of the Cooperative Group Work Rubric to hang in front of the room

Materials

Teacher reference pages

Duplication materials

Extensions

- Addition Extensions 1e and 1h
- Subtraction Extensions 1e and 1h
- Addition and Subtraction Extension
- Multiplication Extension
- Multiplying with Fractions Extension
- Division Extension
- Dividing With Fractions Extension
- Multiplication and Division Extension
- Challenge Extension 1
- Challenge Extension 2
- Challenge Extension 3

Other materials

- Brown paper shopping bags
- Calculators

- Centimeter rulers
- Meter stick
- Oaktag (or heavy-weight paper, 24" x 18")
- Manila file folders
- Marbles
- Pocket folders or large brown envelopes
- Scissors
- Transparent tape
- Masking tape
- Watch/clock with a sweep-second hand or stopwatch
- 2" x 4" x 5" blocks of wood (two per team)
- Plastic/paper bags (for storing track lengths before final assembly)
- $8\frac{1}{2}$" x 11" paper
- Pencils
- Markers

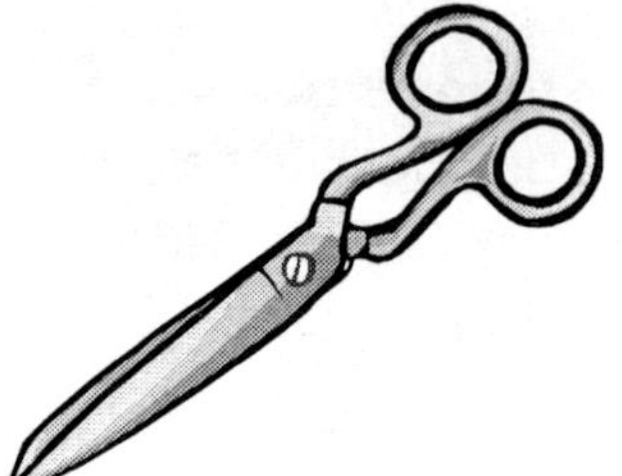

Organizing materials

- You may choose to duplicate all of the materials needed for the unit before starting the unit, or duplicate as you go along. It is a good idea to store each set in its own manila file folder.
- Store the duplication sets in the order they are used. (See Daily Directions.) If a set is used more than once, like the Pretest/Posttest, you may make two separate manila file folders, or just move the folder to its new position after you have used it the first time.

Organizing team folders

- Each team will need a pocket folder or large envelope in which to store its daily work, activity sheets, etc. The Recorder is responsible for the team folder at the end of the day. If there are more materials than the folder can hold, place them in a large brown paper bag, label, and store in a safe place in your classroom.
- Staple two copies of the Score Sheet and one Cooperative Group Work Rubric onto every team folder
- Put one Student Guide into each folder for each team member

Extensions

Fraction Speed Bumps includes a variety of extension activities that enable students to improve their skills while offering teams a chance to earn extra points. Use some or all of the pages for homework challenges or additional classroom assignments in which teams may earn points.

These extension activities may be substituted for other speed bump challenges sheets if you think they will better suit the needs of the class.

- **Addition Extensions 1e** (easy) **and 1h** (hard)—you choose the level of difficulty. Students may work in teams or separately on this challenge.
- **Subtraction Extensions 1e** and **1h**—you choose the level of difficulty. Students may work in teams or separately on this challenge.
- **Addition and Subtraction Extension**
  Students manipulate fractions to reach target answers
- **Multiplication Extension**
  Students look for relationships between factors and common denominators
- **Multiplying With Fractions Extension**
  Students look for patterns
- **Division Extension**
  Students work on solving division problems
- **Dividing With Fractions Extension**
  Students look for patterns
- **Fractions to Decimals Practice Extension**
  Students explore the fraction/decimal relationship
- **Fractions to Decimals Extension**
  Students match decimals to fractions by thinking about the numerator/denominator relationship
- **Multiplication and Division Extension**
  Students manipulate fractions to reach target answers
- **Challenge Extension 1**
  Students use any symbols to get to target answers
- **Challenge Extension 2**
  Students can compete in this game-type challenge
- **Challenge Extension 3**
  Students reflect on previous work as they try to solve the challenge

Assessment

Variety of assessment tools

1. To assess student knowledge of fractions, conduct the Pretest before starting the activities
2. Assess students learning during the unit:
   - Use Individual Assessment sheets keyed to the daily challenges
   - Use additional specialized assessments to spot-check student learning
3. Rubrics will be used to clarify your expectations of students and set a standard of achievement. Use the Cooperative Group Work Rubric throughout the unit to reinforce positive group behavior.
4. Administer the Posttest the day after the Slow Marble Race or on the last day of the unit to assess student understanding of content

Individual Assessments

1. So that you are confident that all of your students are learning new ways of looking at fractions, each challenge sheet has a corresponding Individual Assessment to verify student understanding of the fraction challenge.
2. Each assessment mirrors the type of team-thinking challenge for that day
3. The problems are listed from easy to difficult
4. Students may work all or some of the problems, based on ability and/or teacher expectations
5. Teacher may allow some students to use calculator or get adult help
6. Assessments might also be used to award points (your choice)

Writing—opportunities to clarify math thinking

1. *Fraction Speed Bumps* includes many opportunities for students to write to clarify their math thinking or to demonstrate what they know. You may use these informal assessments as opportunities to earn points.
2. You may ask students to write reflections on different aspects of the unit:
   - Describe a method for quickly solving a problem
   - Describe how they know they have the largest or smallest answer

- Describe what changes they would make in their design
- Write suggestions for changing the rules of the competition
- Evaluate final marble-ramp construction

Peer checking

1. The Checker verifies that the answer of another team is correct
2. When mistakes are found and both teams agree to the mistake, additional points are awarded to the team doing the checking

Focus questions

1. Discussion focus questions are provided to you, which may also be used for written reflections or class discussion

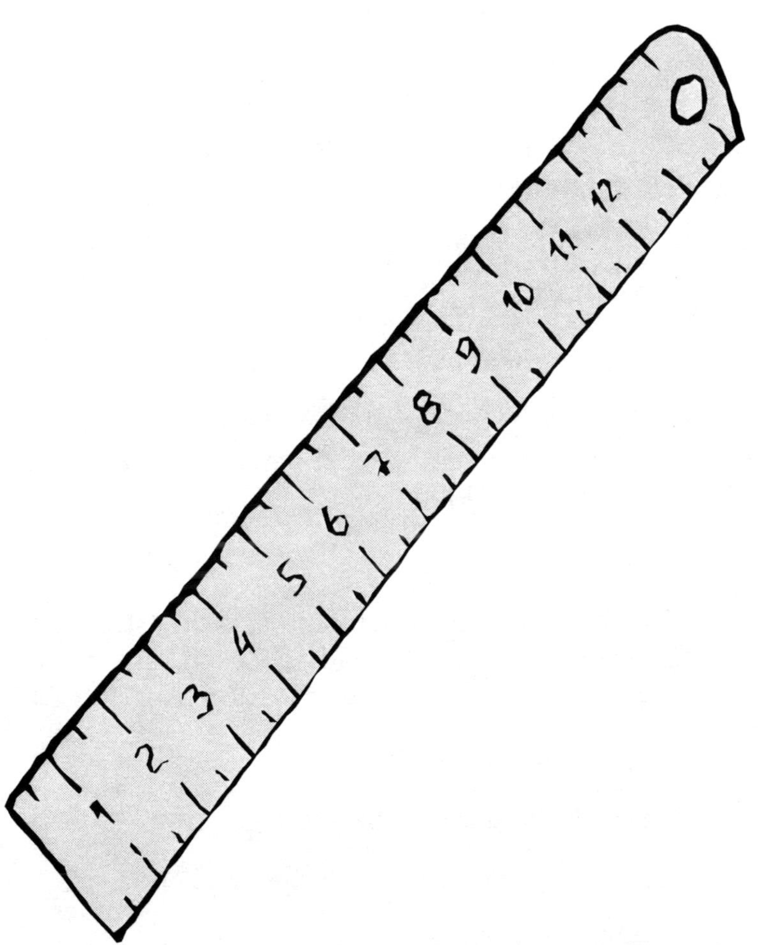

Adapting *Fraction Speed Bumps*

1. Using the fraction speed bump pages

Each page of challenge fraction problems may be used to reinforce a fraction concept. These pages may be used without keeping score and without any of the speed bump design sessions.

a. Use *Fraction Speed Bumps* to supplement your regular classroom instruction. You may begin with any page that matches the lessons you are currently working on.

b. Use the fraction speed bump challenge series at the end of your fraction unit as an enlivening review of many fraction skills

c. Shorten the series by doing only one or two of the activities on each page

2. Optional schedules

Fraction Speed Bumps allows for a competition lasting from one to three weeks based on the challenges selected and the number of competitive events included. The Slow Marble Race will be the culminating activity.

Suggested timeline—The following activities do not have to be held on consecutive days. The unit may stretch over a longer period of time, as you work with your students and introduce the various fraction concepts to them.

- ***13-day experience***

  If students have prior knowledge of most areas of fractions, an intense competition using most of the problems provided could follow this 13-day sequence:

  1. Day 1: Pretest
  2. Day 2: Practice session for familiarizing students with their roles, the daily competition, and how each session will be run. Review the Student Guide. The first session uses only whole numbers.
  3. Day 3: Fractions team task
  4. Day 4: Adding competitions
  5. Day 5: First speed Bump design challenge
  6. Day 6: Subtraction competitions
  7. Day 7: Multiplication competitions
  8. Day 8: Second speed bump design challenge

9. Day 9: Division competitions
10. Day 10: Design session/planning/begin building the Slow Marble Race track
11. Day 11: Finish models and run the Slow Marble Race culmination
12. Day 12: Individual Assessment/Posttest
13. Day 13: Class discussion of or written reflections on the unit

Many of the daily fraction competitions will run longer than one day if you plan to have the teams complete all of the activities. The unit is designed so that you may stop working at any challenge problem during the class period. The next time the teams compete, you might have them begin where they left off, or go on to the next activity. The teams may work on as much or as little as time allows. You determine whether to start a new topic or extend the work on any concept.

- ***6-day experience***

  A **shorter timeline** could run only 6 days, with the teams using only five fraction speed bump pages for the competition. The teacher would have to decide whether to use Pre- and Posttests, which would extend the process to 8 days. In this competition, the teams would earn points to determine a winner, but no culmination would occur.

  1. Day 1: Familiarize students with their roles, the daily competition, and how each session will be run. Review the Student Guide.
  2. Day 2: Fractions team task
  3. Day 3: Adding
  4. Day 4: Subtracting
  5. Day 5: Multiplying
  6. Day 6: Dividing

- ***13-plus-days experience***

  You may want to include reflective writing experiences with this unit. If so, daily writing reflections and discussion questions are included. The unit will then run more than 13 days, but research has shown that writing helps clarify thinking and construct new understandings. An extended experience may increase overall comprehension of these math concepts.

Daily Directions
Day 1

Daily Plan

- Form and name teams
- Introduction, roles and responsibilities (page 2 of the Student Guide)
- Take the Pretest

Materials

- Team folders—*one per group*
- Cooperative work rubric—*one per team (or class chart)*
- Score sheet—*one per team*
- Student Guide—*class set*
- Pretest/Posttest—*class set*
- Pencils
- Markers

Teacher Reference

- Pretest/Posttest Key

Preparation

Small group

1. Assign students to heterogeneous teams and set up team folders. See setup directions on page 16.
2. Designate a work area and an area for each team to store all materials. Three desks side-by-side make for the best work area.

Procedure

1. Announce team assignments and allow teams a few minutes to select team names
2. Distribute team folders and direct students to write their team name on the outside of the folder. Under their team name, they should write their individual names.

   a. Teams need their folders every day

   b. The Recorder is responsible for the folder during class time and keeps the materials organized

   c. Team folders are turned in at the end of each class period

4. Distribute individual Student Guides and have each student write the team name and his/her own name at the top. Read the roles and responsibilities on page 2 of the Student Guide.

5. Answer any questions and collect team folders

6. Administer the Pretest and correct it before the class meets again

Individual

Daily Directions
Day 2

Small group

Daily Plan

- Assign roles of Setter, Recorder, and Checker
- Review Student Guide and practice teamwork
- Work on team practice session

Materials

- Student Guide—*one per student*
- Teamwork Practice sheet—*class set*
- Score sheet—*one per team*
- Calculators

Teaching tip
This page may be skipped if you want to start with the fraction speed bump sheet from day 3. This activity could be used later for earning additional points.

Teacher Reference

- Student Guide
- Teamwork Practice Key

Preparation

If the pretest reveals that your students need a lot of instruction, put together materials to supplement the information in the Student Guide.

Procedure

1. Distribute team folders and assign roles. The roles will rotate each time the team competes, but each role may not have a specific task each day. The Recorder rotates to become the Setter, who rotates to become the Checker, who rotates to become the Recorder.
2. Assign each team an opponent for checking purposes. (See identifying opponents in the About the Unit section.)
3. Have students open their Student Guides and read the numbered headings of the fraction review topics.
4. Review the first two informational items.
5. Use the Teamwork Practice Session sheet to give each team practice. **Note:** *This page uses only whole numbers to get students used to the procedures they use throughout the unit.*

6. Discuss **Challenge 1**. Have students write their team name on the top of the paper. Have the students fold the paper so that only the first task is visible. They are trying to get the highest score they can by planning and thinking about relationships. Encourage the teams to have each person work on a solution. Allow the teams time to discuss their strategies and how to best use teamwork to attack the first challenge.
7. Ask if there are any questions. For this activity, have the Recorder sit between the other two students.
8. Allow them to use calculators for these challenges because the focus is teamwork, not fractions
9. Let them work for a while and then say, "Only one minute left. Please finish by filling in the boxes and writing your best answer in the answer space."
10. Their final highest score becomes the point value they earn for each challenge. They can earn ten bonus points for a whole-number answer. **Work on the second and third challenges as time permits.** This activity will help you see how the teams are working out.
11. When the competition is over, the Checker must double-check the work of the opposing team. He/she may use a calculator for speed and accuracy in checking. Direct the students to "How to Fill Out the Score Sheet" in the Student Guide. The Recorder enters the score and any bonus points for this first session. The other team members will become the Recorder in other sessions.
12. Put all papers in the team folder along with the Student Guides
13. Discuss strategies for teams succeeding, focusing on using all members' input

Note: *If the upcoming Fraction Speed Bump Challenges for each day appear to be too challenging, please look at the Extensions section of the unit. Many of these activities can replace a difficult challenge on any topic and still offer challenges to most students. Just follow the unit's schedule, but use the simple instructions from the Extension page instead. You may need to modify how points are awarded, but there are plenty of scoring suggestions in the unit for you to adapt. This flexibility allows you to run the competition at a challenge level that works for you and your class. Have fun.*

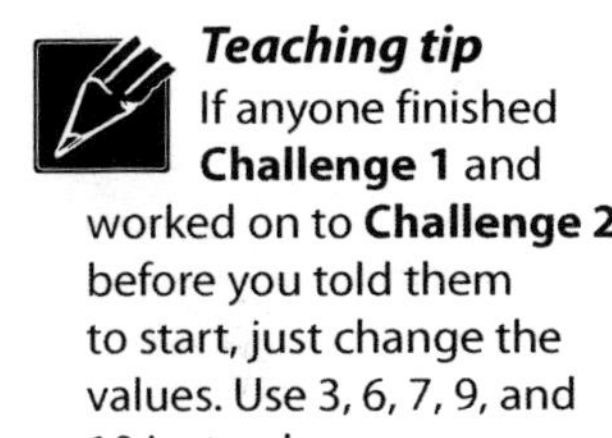

Teaching tip
If anyone finished **Challenge 1** and worked on to **Challenge 2** before you told them to start, just change the values. Use 3, 6, 7, 9, and 10 instead.

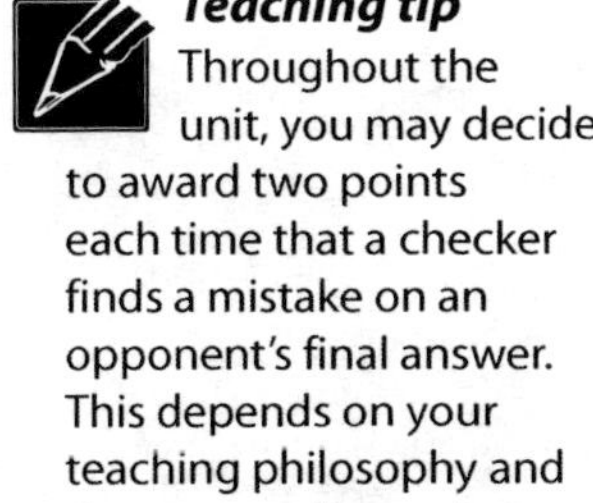

Teaching tip
Throughout the unit, you may decide to award two points each time that a checker finds a mistake on an opponent's final answer. This depends on your teaching philosophy and the nature of your class.

Daily Directions
Day 3

Small group

Daily Plan

- Review Student Guide information with students
- # 1—What is a fraction?
- #2—Every fraction is just a division problem
- #3—Every whole number can be changed to a fraction
- #17—Least common multiple
- #18—Greatest common factor
- Fractions Speed Bump: Day 3

Materials

- Fractions Speed Bump: Day 3—*one per team (optional—whole class set)*
- Scrap paper for each person
- Calculators—*one per team (optional)*

Teacher Reference

- Fractions Speed Bump: Day 3 key

Procedure

1. Distribute folders and rotate roles. The Recorder rotates to become the Setter, who rotates to become the Checker, who rotates to become the new Recorder. Put teams in three-desk working areas with the new Recorder in the middle.
2. Discuss working together, sharing information with teammates, double-checking work and writing the team's final answers neatly in the boxes
3. Tell each team who their opponent is for the day
4. Review concepts #1, #2, #3, #17 and #18 from the Student Guide
5. You may choose to write examples of equivalent fractions on the board before the teams start

6. Allow the teams time to discuss their strategies and how to best use teamwork to attack the first challenge
7. Pass out one Speed Bump: Day 3 sheet (face down) to each team
8. Tell teams to turn papers over and put team name on the paper and then begin with **Challenge 1**
9. When a team has finished, it may raise a hand; you may check for correctness and put a "10" on the first sheet if it has the correct answers. If incorrect, urge them to try again. Place a "9" on second place, an "8" on third, etc. The first place team earns 10 points, second place—9, third—8, fourth—7, fifth—6 and all other teams earn 5 points.
10. Since some teams may finish quickly, you can stop the competition after half of the teams finish task one. All teams not finished, earn 5 points.
11. You write the answers on the board and Checkers confirm that the numbers are correct
12. Discuss the strategies that teams used to succeed
13. Assign one or both of the remaining challenges based on the time remaining
14. Write the new values for **Challenge 2** on the board to start the competition. **The values for the second challenge are 2, 4, 6, 8, 10, 12, and 14. The values for the third challenge are 1, 1, 2, 2, 2, 4, 7, and 20.**
15. As the teams finish each challenge, write the points they earn (10, 9, 8, etc.) on their sheets
16. Have the Recorder fill out the team score sheet and place it in the team folder

Discussion/Written Reflection Questions

Challenge 2

1. Why must the first denominator box be a 6?
   It is a factor of all the other denominators. The 2 is, but can't be placed there because it creates an improper fraction.
2. Why must all the numerators be even numbers?
   You only have even numbers.
3. What numbers must go in numerator boxes 2 through 5? Why?
   8, 10, 12, 14. Since the fractions are all equivalent, as the denominators increase, the numerators must increase.

Challenge 3

1. Why is 20 the only number that can go in the last denominator box?
   It is half of 40; 11 is half of 22. The 5 can't go there because you would divide 40 by 8 to get that denominator, but you can't divide 22 by 8.

Written reflections *(for bonus points)*

These focus questions may be used at different points throughout the unit. They may not be appropriate for this first activity, but they should be kept in mind throughout the unit.

1. Have team members all write why/how they know they have the highest or lowest answer to a challenge
2. Have team members write what method they used to solve a challenge quickly

Daily Directions
Day 4

Daily Plan

- Adding Fractions speed bump
- Review Student Guide information with students
- #4—You may only add or subtract fractions with like denominators
- #5—If the labels are different, when you add or subtract, you must change one or both in order to have the same labels
- Individual Assessment: Addition

Small group

Materials

- Adding Fractions Speed Bump: Day 4
- Calculators—*one per team (optional)*

Teacher Reference

- Adding Fractions Speed Bump: Day 4 Key
- Individual Assessment: Addition Key

Procedure

1. Put teams in three-desk work areas and rotate roles
2. Distribute team folders
3. Direct students to Student Guide to review #4 and #5
4. You may choose to review a few addition problems
5. Review the definitions for proper and improper fractions found in the Student Guide
6. Allow the teams time to discuss their strategies and how to best use teamwork to attack the first challenge
7. Pass out Speed Bump Day 4 sheet (face down) to each team
8. **Challenge 1**—Score as before (i.e., 1st, 2nd, and 3rd get 10, 9, 8, etc.)
9. Write the answers on the board and have Checkers confirm that the numbers are correct

Teaching tip
If some teams are having trouble, you may choose to place a value in a box for them when you pass their work area. Example: Place a 5 in the box above the 6 making, the fraction $\frac{5}{6}$. This reduces the number of possibilities for the remaining values.

Teaching tip
Since it is addition, the addends may be in a different order, but will still be correct.

10. If the first two challenges are too easy for your class, then only award points to the first two teams that finish correctly. You may simply skip the first two challenges and move to the more challenging activity #3.
11. The Checker checks an opponent's work to verify that it is correct
12. Teams earn ten bonus points for problem three if they find the largest answer
13. If some teams do not finish, they can earn one point for each value written in the correct box
14. After Challenge 1 is completed, you may write the values to **Challenge 2** on the board. **The values for Challenge 2 are 1, 2, 4, 5, 6, and 12.** For challenges two and three score the same way as before (1st, 2nd, and 3rd get 10, 9, 8, etc.)

Later, write the values to Challenge 3 on the board if you choose to include Challenge 3. **The values for Challenge 3 are 1, 2, 4, 5, 6, 12, and 25.**

1. Write the values for **Challenge 4** on the board when needed. **The values are 2, 3, 4, 5, 6, and 7**
2. **Challenge 4**—Score as in step 14 (above)
3. Teams earn an additional 7 points if they have the correct largest answer. A team can also earn an additional 7 points for having the smallest answer. Other correctly worked problems without the largest or smallest answer get 3 points. Checkers must verify that the problem was written down and solved correctly.

Discussion/Written Reflection Questions

Challenge 1

1. Where is a natural starting point for solving this challenge?
   The box to the right of the fraction $\frac{20}{12}$.
2. Are there any numbers that cannot go in any of the denominator boxes?
   The number one. In the answer, none will work except the number three.

Challenge 3

1. What math strategy did you employ to generate the largest answer?
   Work backwards and place the 25 and the 12, and then make the three largest fractions with the values that remain.

Challenge 4

1. How would you get the largest answer if you could use improper fractions for this challenge? Match the smallest denominator with the largest numerator.
   $\frac{7}{2} + \frac{6}{3} + \frac{5}{4}$

Daily Directions
Day 5

Small group

Daily Plan

- Participate in first speed bump event

Materials

- Speed bump templates
- $8\frac{1}{2}$" x 11" paper (for taping down speed bumps)
- Tape
- Marbles (more than 6)
- Meter stick
- Masking tape
- Table or two desks that are the same height

Teacher Reference

- Speed bump information, pages 52–60

Preparation

1. Prepare the speed bumps for the option chosen (one or two crooked speed bumps or the three, single speed bumps)
2. Cut, fold, and tape down the speed bumps on $8\frac{1}{2}$" x 11" paper before class starts
3. Cut, fold, and tape the catching tray (following the instructions) before class starts
4. Divide the tray into four sections and write the fractions ($\frac{1}{10}$, $\frac{1}{6}$, $\frac{1}{5}$, $\frac{1}{4}$) in the four areas. (You may choose other fractions.)
5. Mark the table with masking tape. Put a six-inch line of tape at the starting line. Put a piece of tape at the 50 cm mark and another piece at the 100 cm mark.
6. The speed bump $8\frac{1}{2}$" x 11" paper will be taped at the 50 cm mark
7. Tape the front of the catching tray at the 100 cm mark

8. Place the meter stick behind the catching tray to stop most marbles from rolling onto the floor

9. Designate a person to gather marbles that reach the floor

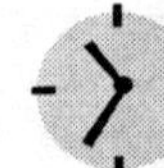
All three options take 15 minutes

Options

1. Provide one crooked speed bump. No team points used and teams can earn points.

2. Provide two crooked speed bumps. Teams must spend 4 or 8 points to purchase a speed bump for their opponent.

3. Provide three premade speed bump sheets. Purchase one of the three choices of speed bumps (small, medium, or large) for your opponent.

Procedure

1. Pass out team folders

2. Assign opponents for the speed bump marble-rolling event

3. Show the premade speed bumps for the option(s) you have chosen, and tell how much they cost. (Crooked speed bumps cost 4 and 8 points. The small bump costs 5 points, the medium 7 points, and the large 3 points.)

4. Teams need to decide which speed bump to purchase to slow down their opponent

5. Teams enter points spent on speed bumps on their score sheets

6. Tape down the first speed bump sheet on the table at the 50 cm mark

7. All teams that must use the first speed bump sheet, roll first

8. Each team has one member who rolls all six marbles

9. The Recorder is the first team member to roll the marbles. The Roller position rotates with each marble-rolling event. The Recorder (then Setter, and then Checker) should release the marble from a six-inch starting line 100 cm from the front of the catching tray.

10. The marble must be released along the six-inch starting line

11. When a marble lands in the tray on any fraction, the marble is removed, and the team member continues to roll. Once a marble lands on the largest fraction area, the team can stop rolling. No more points can be earned. The largest fraction earned is then used to compute the team's points for this activity.

Teaching tip
Place the table where most students can watch, or allow the students to move closer to the table.

Bright Idea
If you do not have a table, you can use two desks and tape the speed bump paper over the crack.

Teaching tip
Each student who rolls only gets one minute to roll the six marbles into the catching tray.

12. The Recorder multiplies the fraction by the current point total for the team and adds that answer (rounded off) to get a new higher total
13. The opponent's Checker verifies the math and allows the "points earned" to be written on the correct score sheet
14. Tape down other speed bump sheets as needed, so all teams get to roll marbles over the speed bump
15. Continue recording scores and placing the points earned on each team's score sheet
16. Collect all materials

Teaching tip

If you choose to use **only** the premade crooked speed bumps **or** only the single speed bumps the marble rolling part should only take 15–20 minutes. Using both options will better fill a 45 minute class period.

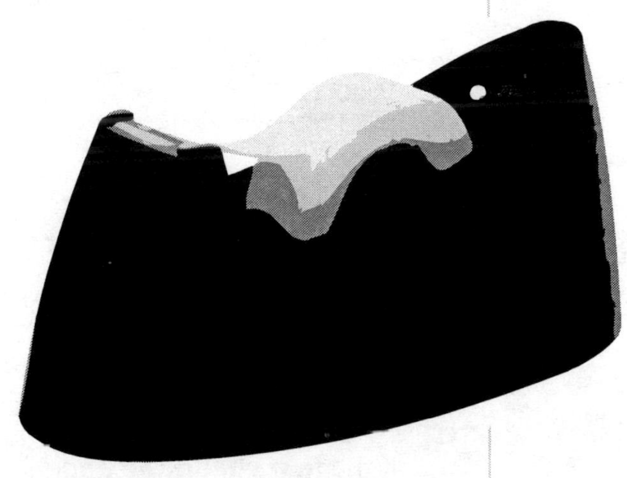

Daily Directions
Day 6

Small group

Daily Plan

- Subtracting Fractions speed bump
- Review Student Guide information with students
- #4—You may only add or subtract fractions with like denominators
- #5—If the labels are different, when you add or subtract, you must change one or both so that they have the same labels
- #6—In a mixed numeral subtraction problem, you may need to regroup
- Individual Assessment: Subtraction

Materials

- Subtracting Fractions Speed Bump: Day 6
- Calculators—*one per team (optional)*
- Scrap paper

Teacher Reference

- Subtracting Fractions Speed Bump: Day 6 Key
- Individual Assessment: Subtraction Key

Procedure

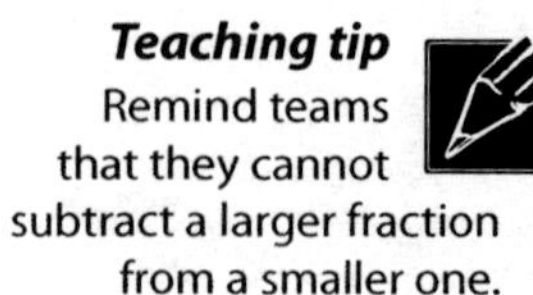

Teaching tip
Remind teams that they cannot subtract a larger fraction from a smaller one.

1. Put teams in three-desk work areas and rotate roles
2. Distribute team folders
3. Direct students to Student Guide to review #4, #5, and #6
4. You may choose to review subtracting fractions
5. Allow the teams time to discuss their strategies and how to best use teamwork to attack the first challenge
6. If the **first challenge** appears to be too difficult for your students, tell them one of the two fractions and then let them work on the challenge. Example: Write $\frac{2}{20}$ in the boxes to make the fraction that you will be subtracting.

7. Allow a reasonable amount of time for problem solving, then direct the Recorders to enter each team's best effort
8. Have Checkers verify the opponent's answer
9. Share largest answer from the teacher's key and award points
10. Teams with the highest answer earn ten points. Teams with the second highest answer earn six points. All other answers earn four points for trying.
11. Teams record scores and bonus points (6) on their score sheets
12. **Second challenge**—Write the values **1, 2, 3, 4, 5, and 6** on the board and allow the teams to work on Challenge 2. Teams with the lowest answer earn ten points. Teams with the next-to-lowest answer earn six points. All other teams earn four points for trying.
13. **Third challenge**- Write the values **1, 2, 4, 8, 10, and 12** on the board when needed.
14. Teams with the highest numerical answer earn ten points. Teams with the second highest answer earn six points. All other answers earn four points for trying.

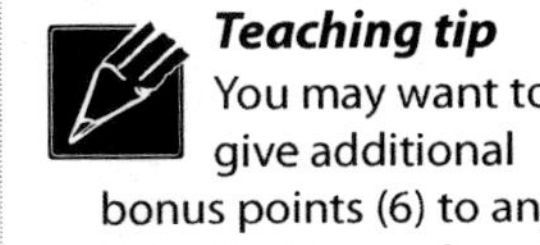

Teaching tip
You may want to give additional bonus points (6) to any team that has reduced the final answer to lowest terms.

Discussion/written reflection questions

First challenge

1. What should we try to create in the first fraction boxes?
   The largest fraction we can.

Second challenge

1. What strategy did you use in searching for the smallest answer?

Third challenge

1. What strategy is used in solving this challenge?
   Create the largest fraction possible to add in the second part of the challenge. Use the other four values to create the largest answer when subtracting in the first part.

Daily Directions
Day 7

Small group

Daily Plan

- Multiplying Fractions speed bump
- Review Student Guide information with students
- #7—In a word problem, "of" means ***multiply*** and "is" means ***equal to***
- #8—When you multiply any value using a proper fraction, your answer will be a smaller part of the original value
- #13—Multiplication is the exact opposite of division, and division is the exact opposite of multiplication
- #14—Reduce fractions when you can before beginning to solve a problem
- Individual Assessment: Multiplication

Materials

- Multiplying Fractions Speed Bump: Day 7
- Calculators—*one per team (optional)*
- Scrap paper

Teacher Reference

- Multiplying Fractions Speed Bump: Day 7 Key
- Individual Assessment: Multiplication Key

Procedure

1. Put teams in three-desk work areas and rotate roles
2. Distribute team folders
3. Direct students to Student Guide to review #7, 8, 13, and 14
4. You may decide to review some multiplication problems
5. If the **first challenge** is too difficult for your class, you can discuss ways to attack the problem and then award points if someone suggests a strategy for how to begin or offers values that are correct. Allow all teams to work from that helpful start.

6. Allow the teams time to discuss their strategies and how to best use teamwork to attack the first challenge
7. Allow a reasonable amount of time for problem solving, then direct the Recorder to enter the team's best effort
8. Remind the teams to circle the word that indicates whether they are trying for the answer with the highest or the lowest numerical value
9. Have Checkers verify the opponent's answer
10. Share the largest and then smallest answer from the answer key, and award points
11. If only one team has the highest answer, it earns ten points. If two teams have the same highest answer, each gets 9 points. Three teams tied get eight points each etc. Use the same method when awarding points for lowest answer. ***Bonus:*** If only one team selects highest or lowest and has the correct answer, it gets seven additional points. All other team answers earn five points for trying. ***Remember: teams can compare their fraction answers by using calculators and dividing to get decimal answers.***
12. **Second challenge**—Allow teams a minute or two to discuss strategies before placing the values on their opponent's paper. The Setter writes the values in when you say it is time to start. All teams that have the correct high answer earn 10 points. Any teams that tie for the second highest answer get 8 points. All others get 5 points.
13. **Third challenge**—Allow teams a minute or two to discuss strategies for placing the values on their opponent's paper. The Setter writes the values in when you say it is time to start. A team with a higher answer than its opponent earns 10 points. The opponent gets 4 points. Any one team with the highest answer in the class gets 15 bonus points.
14. Record points earned and bonus points on score sheet

Teaching tip
To get the largest answer, make the three largest fractions possible and then multiply. The fractions may be arranged in any order.

Teaching tip
To get the lowest answer, make the three smallest fractions possible and then multiply.

Daily Directions
Day 8

Small group

Daily Plan

- Participate in speed bump design and/or speed-bump-trap design competition

Materials

- Speed Bump Template
- Speed Bump Trap Template
- $8\frac{1}{2}$" x 11" paper (for taping speed bumps to)
- Tape
- Marbles
- Meter stick
- Masking tape
- Table or two desks that are the same height

Teacher Reference

- Speed bump information (pages 52–60)

Preparation

For speed bumps:

1. Print one Speed Bump Trap Template for each team.
2. Give each team one $8\frac{1}{2}$" x 11" sheet of paper
3. Mark the table with masking tape, as before, to speed placement of speed bumps, indicate the six-inch starting line, and mark where the "catching tray" is to be taped down (100 cm from the starting line)
4. Have catching tray folded and taped to the table for team-made speed bumps option
5. Have speed bump trap example folded and taped to show the class
6. Place the meter stick behind the catching tray to stop most marbles from rolling onto the floor
7. Designate a person to gather marbles that reach the floor

For speed bump trap:

1. Print one Speed Bump Template for each team
2. Print one Speed Bump Trap Template for each team

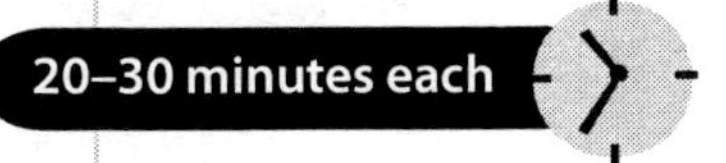

Options:

1. Team-made speed bumps: Teams purchase up to two speed bumps, cut out, and tape on an $8\frac{1}{2}$" x 11" sheet of paper as obstacles.
2. Team-made speed bump trap: Teams purchase any number of speed bumps and create a speed bump trap.

Option 1—Team-made speed bumps

Procedure:

1. Teams may buy up to two speed bumps (small, medium, or large), in any combination
2. A small bump cost five points, a medium bump costs seven points, and a large bump costs three points
3. Teams enter points spent for speed bumps on their score sheet
4. A team may choose to spend no points and let its opponent roll at the tray with no speed bumps in the way
5. Having purchased its speed bumps, the team decides how it wants to place them on an $8\frac{1}{2}$" x 11" sheet of paper and then secretly tape them down. All teams must cut out, fold, and then tape the speed bumps down; no team may change its speed bumps. Any configuration is allowed, but the speed bump edges must be taped flat, using the required minimum spacing (see Speed Bump Template), and may not touch.

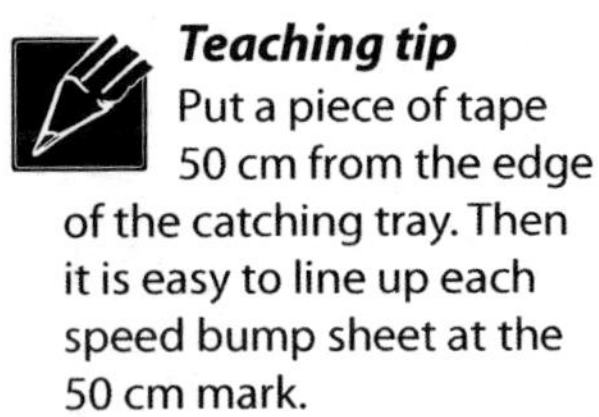

Teaching tip
Put a piece of tape 50 cm from the edge of the catching tray. Then it is easy to line up each speed bump sheet at the 50 cm mark.

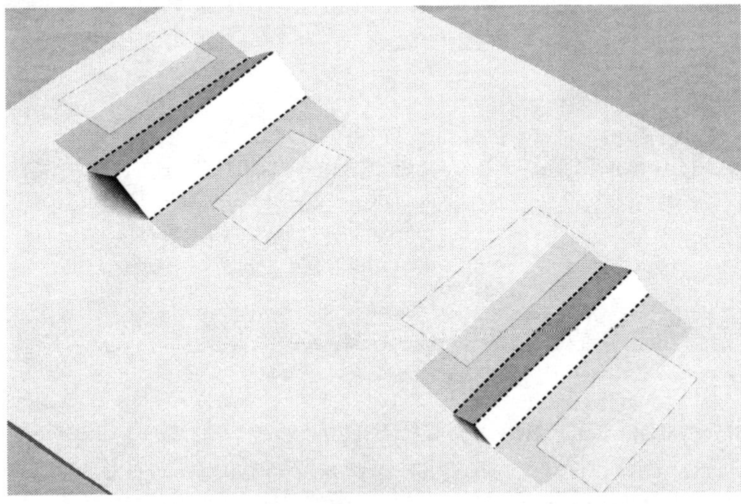

Speed bumps taped down to paper.

Teaching tip
Tell the rollers that they only have one minute to roll all six marbles.

6. The speed bump sheet is placed 50 cm from the front of the catching tray
7. The Roller (the position rotates with each marble-rolling event: first the Recorder, then Setter, then Checker) should release the marble from a six-inch starting line 100 cm from the front of the catching tray
8. Mark the starting line with six inches of masking tape. All marbles should be released somewhere on that line of tape to prevent Rollers from taking an angle that always misses the speed bumps.
9. Decide how many "rolls" each team gets (we suggest 6)
10. Each team has one member roll all six marbles
11. If this is the fourth time teams have rolled marbles over speed bumps, then each team has the option to spend 5 points and choose who will roll the marbles, no longer following the designated rotation established earlier
12. When a marble lands on a fraction, the marble is removed, and the team member continues to roll. Once a marble lands on the largest fraction, the team can stop rolling. No more points can be earned. The largest fraction earned is then used to compute the points for this activity.
13. Marbles that miss the tray earn nothing, but if a team lands more than one marble on the tray, it uses the largest fraction to compute its bonus points. If a marble lands on a line, the team gets the higher fraction.
14. The Recorder multiplies the fraction by the current point total for the team and adds that answer (rounded off) to get a new higher total
15. The opponent's Checker verifies the math and okays the "points earned" on the correct score sheet
16. Collect all materials
17. ***Option:*** You could have a few problems for the team to work correctly before they can record their newly earned points
18. Eight teams rolling six marbles takes less than 15 minutes

Option 2—Team-made speed bump trap (six rolls)

Procedure:

1. A speed bump trap may be created instead of using a speed bump sheet. The speed bump trap incorporates the speed bumps and catching tray into one unit.

2. Any number of speed bumps may be purchased at their regular point price. Teams cut out, fold, and then tape down the speed bumps in the speed bump trap that they have assembled. The speed bumps may not touch each other when taped down.

3. Teams enter points spent for speed bumps on their score sheet

4. The team must divide the tray into four separate sections with approximately the same area in each

5. The teams write a different fraction in each area. Since the fraction becomes the multiplier for earning points, the teacher must offer the four fractions that all teams use. ($\frac{1}{10}$, $\frac{1}{5}$, $\frac{1}{4}$, and $\frac{1}{6}$ are examples.)

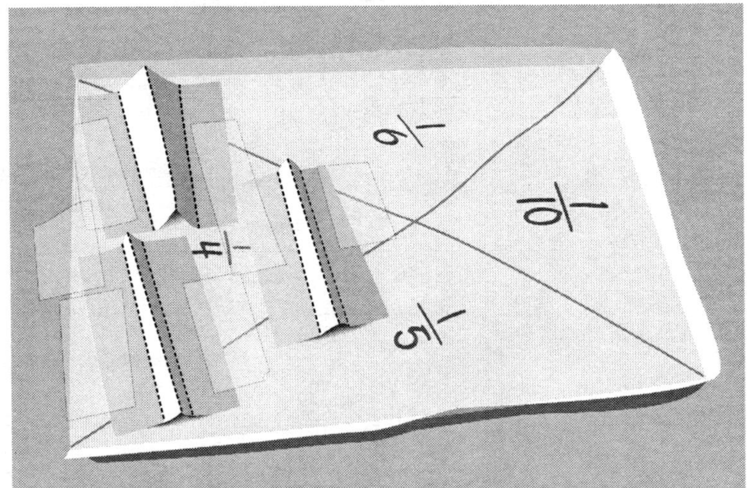

Speed bumps taped down to a speed bump trap.

6. A team may skip using speed bumps and instead divide the trap into four areas with a fraction written in each area

7. The speed bump trap is taped to the table with the front edge 100 cm from the starting line

8. The opponent's team has one member roll all six marbles

9. The Roller (position rotates with each marble rolling event) should release the marble from a starting line 100 cm from the front of the speed bump trap

10. Marbles that miss the tray earn nothing, but if a team lands more than one marble on the tray, it uses the largest fraction to compute its bonus points. If a marble lands on a line, the team gets the higher fraction.

11. When a marble lands on a fraction, the marble is removed, and the team member continues to roll

12. The Recorder multiplies the fraction times the current point total for the team and adds that answer (rounded off) to get a new higher total

13. The opponent's Checker verifies the math and okays the "points earned" on the correct score sheet

14. Collect all materials

Daily Directions
Day 9

Small group

Daily Plan

- Dividing Fractions speed bump
- Review Student Guide information with students
- #9—Dividing is just asking this simple question: "How many of these are in that?"
- Individual Assessment: Division

Materials

- Dividing Fractions Speed Bump: Day 9
- Calculators—*one per team (optional)*
- Scrap paper

Teacher Reference

- Dividing Fractions Speed Bump: Day 9 Key
- Individual Assessment: Division Key

Procedure

1. Put teams in three-desk work areas and rotate roles
2. Distribute team folders
3. Direct students to Student Guide to review #9
4. You may decide to review division of fractions
5. Allow the teams time to discuss their strategies and how to best use teamwork to attack the first challenge
6. **First challenge**—Pass out Dividing Fractions Speed Bump: Day 9. Allow a reasonable amount of time for the class to work on these challenges. If possible, stop the class when a few teams have finished the bonus. Direct each Recorder to enter their team's best effort.
7. **Points—**10 points for all correct answers; 3 for all other correctly worked answers. Bonus is worth 5 points.
8. Have Checkers verify the opponent's answers and award points

Teaching tip

If the students struggle, point out that if you want to get a large answer make the first fraction (dividend) large and the second fraction (divisor) small. Example: $\frac{8}{2}$ divided by $\frac{4}{6}$.

9. ***Remember: Teams can compare their fraction answers by using calculators and dividing to get decimal answers.***

10. **Second challenge**—Write the values **1, 3, 5, 7, 8, and 10** on the board and allow the teams to begin the second challenge

11. All teams that have the smallest answer earn 10 points. Any team that competes for the smallest answer but does not have it, loses 6 points. If only one team circles "smallest" and has the correct answer, that team earns 15 points instead of 10.

12. Do not pass out the second speed bump sheet (third challenge) until it is needed

13. **Third challenge**—Allow teams a minute or two to discuss strategies for placing the symbols on their opponent's paper. The Setter writes the symbols in when you say it is time to start. All highest answers get 10 points; all lowest answers get 8 points. If many teams have the correct answer, they all get the full point total. Teams that do not have the highest or lowest answer earn 2 points if they have a correctly worked problem.

14. Teams record all points, penalties, and bonus points on their score sheets

Discussion/Written Reflection Questions

Challenge 1

1. How many division problems are possible?
   4

2. In the second part of the challenge, how do improper fractions help you get a larger answer?
   You can create a whole number and divide it by a smaller fraction.

Challenge 2

1. How does a team know it has the smallest answer?
   By taking the smallest fraction and dividing that by the largest fraction and then dividing that answer by the remaining fraction.

Challenge 3

1. Which combination of symbols is best to get a large answer?
   Add the largest two fractions and then divide by the remaining fraction.

2. Is there a way to place symbols so that you will beat your opponent?
   It may depend on where the opponent places your symbols. "Multiply, then subtract" is a good strategy to keep your opponent's score low. To win you need a high or a low total. Maybe add then subtract (or subtract then add) will keep your opponent from getting the largest or smallest.

Daily Directions
Days 10 and 11

Small group

Daily Plan

- Design, build, and test Slow Marble track
- Slow Marble Race information (Student Guide)

Materials

- Team folders
- Oaktag or heavy paper
- Scissors
- Tape
- Stopwatch or a clock with a sweep-second hand
- A volunteer to time all teams' marble rolls
- Centimeter ruler
- Marbles—*one per team*
- Calculators—*one per team (optional)*
- 2" x 4" blocks of wood—*two per team*
- Plastic or paper bags to store track sections and pillars—*one per team*
- Markers to write team names on storage bags

Information

Use the following information to prepare for the Slow Marble Race.

Teaching tip
Determine how much track length each team gets before the start of this session.

1. Students are awarded centimeters of track based on their total points earned. See the Slow Marble Race overview on page 61.
2. The best track-design strategy is to use the materials earned and create a track that slows the marble's descent.
3. Planning done quietly and secretively will give teams an advantage, as no other team will see or hear their ideas and get to use them
4. Allow the teams to spread out to the far corners of the room to keep their planning secret

5. Encourage cutting, folding, testing and modifying of ideas, and initial construction on the first day. Teams must use materials carefully during testing. Once started, there is no way to get more building supplies for additional track.

6. All teams use two blocks of wood to anchor the track at the top end. Tape the track to the top block of wood. This assures everyone that they all have the same starting point (approximately 3 inches from the floor).

Teaching tip

You need to make the final competition fair. If one or two teams have so many more points than the other teams, you may declare them the winners of the first section of *Fraction Speed Bumps*. You then can award track lengths based on team totals: 1st place gets 210 cm, 2nd gets 205, 3rd gets 200, etc. Alternately, you can choose to award every team the same amount of track (200 cm), giving every team an equal chance at success. Your decision depends on the class and what you are trying to achieve.

7. Each team needs about 200 cm of track to compete. If all teams have less than 200 cm, you may tell them that they all receive the same additional number of points for working well during the competition. You want your middle team to have about 200 cm of track.

8. On the day of the Slow Marble Race finals (Day 11), teams finish building their racetrack and get in as many timed test-runs as possible. They make modifications to the track after each test run in an attempt to increase their times.

9. Teams that cooperate will finish building quickly and have more time to test and refine their racetrack. They will also be able to have more trial runs tested than other teams.

10. Set an ending time so the class knows when the competition is over

11. After the first day of racetrack preparation, one class period is enough time for a team to construct, modify, and time a number of test runs

12. The final competition is ongoing. Each team tests its racetrack as often as possible, and when ready, has a timer record its time on the board. They can continue to modify the track and try to increase their time. A timer will record each new "slower" time on the board next to the team's name.

13. Since all times are written on the board, teams can see how much they need to improve to win

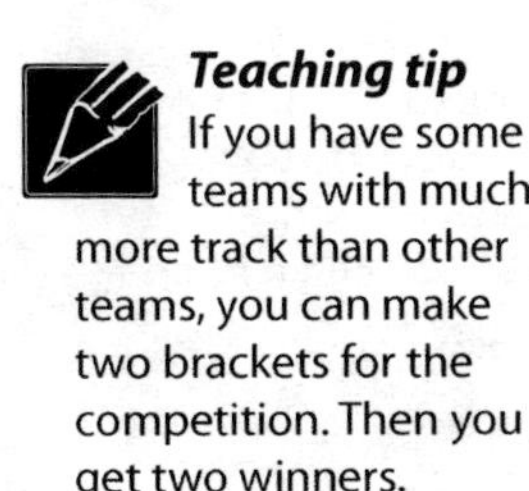

Teaching tip

If you have some teams with much more track than other teams, you can make two brackets for the competition. Then you get two winners.

Procedure for First Day

First day of designing and construction:

1. Distribute team folders

2. Prepare samples of the race track and a folded corner to share with the class

3. Direct students to the folding instructions for the race track and the support pillars in the Student Guide. The teacher may want to demonstrate each of the folding techniques to the class.

4. If necessary, demonstrate how to draw the 2 cm parallel lines for the track and how to cut them out. Demonstrate how to fold the U-shaped track. Demonstrate how to cut and fold a corner.

5. Tell each team its total track length. Explain the method you used to determine their totals.

6. Distribute heavy paper and tape to each team

7. Distribute some $8\frac{1}{2}$" x 11" paper (for drawing their racetrack designs)

8. Distribute two, 2" x 4" blocks of wood to each team. These blocks of wood determine the high point of the racetrack.

9. Review the Cooperative Group Work Rubric and encourage everyone to work for the team. While one member is drawing parallel lines, another might be designing the track. Shortly after, one can cut the track so another can fold it.

10. An opponent's Checker then verifies that the other team has cut out the correct length of track. You may want to take on this responsibility yourself.

11. Make sure each team has a place to work on the floor

12. Tell teams to begin

13. Move around the room answering questions and checking that all team members are contributing

14. Remind them to tape the top of their track to the top block of wood

15. Tell them that they are allowed to tape support pillars to the floor

16. Make sure they tape the pillars to the side of the track

17. Stop them with enough time remaining so that they can place all items in a bag when cleaning up for the day

18. Have them write their team name on the bag and place all items inside

19. Put the bags in a safe place until the final construction day

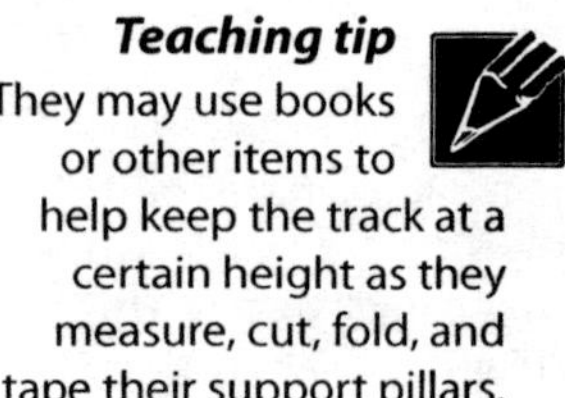

Teaching tip
They may use books or other items to help keep the track at a certain height as they measure, cut, fold, and tape their support pillars.

Procedure for second day

Second day of designing, final construction, and testing:

Teaching tip
Enlist another adult to help you with the timing of the marbles.

1. Distribute all materials in team storage bags

2. Place team names on the board and record their times for all to see

3. Identify ending time for last timed runs

4. Have teams begin
5. Move around the room, encouraging teams and timing their marble runs
6. Frequently announce how much time is remaining in the competition
7. Keep writing the slower times on the board by each team's name
8. Stop the event with enough time to announce the winner(s) and put away all materials

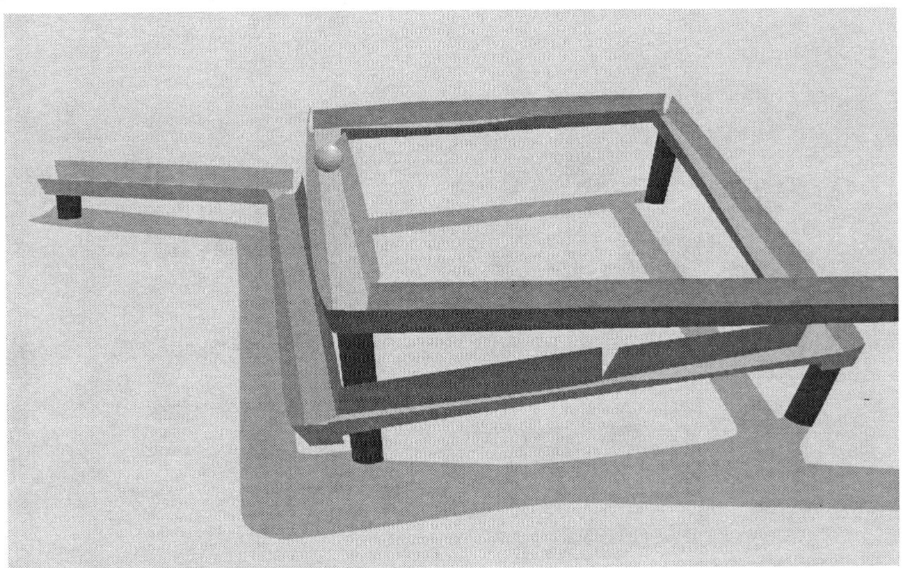

Part of a finished marble track.

Daily Directions
Day 12

Small group

Daily Plan

- Class discussion

Materials

- List of discussion questions

Teacher Reference

Procedure

1. Pass out the storage bags and ask teams to quickly assemble their tracks
2. **Discussion questions:**
   a. Which designs worked best, and why?
   b. Were any designs unique?
   c. Are there ways to slow the marble down that we have not tried yet?
   d. Are there other materials that should be included in this race?
   e. Why didn't we start the race at 12 inches, instead of 3?
   f. What design element worked best in slowing the marbles?
   g. Can we use speed bumps in this race? How?
   h. What if the track was a one-inch "U," instead of a 2 cm "U"? Would that open up other possibilities for slowing down the marbles?
9. Encourage students to continue using their creativity to solve problems
10. If interested, the teams can redesign their tracks and compete again on another day. This time, you could open up the event and allow them to bring in any item from home and use it in an attempt to create the slowest descent on the track they have.
11. Collect all materials
12. Assign written reflections as appropriate

Daily Directions
Day 13

Individual

Daily Plan

- Take the Posttest

Materials

- Pretest/Posttest

Teacher Reference

- Pretest/Posttest Key

Procedure

1. Pass out the Posttest
2. Encourage students to reflect on the skills and strategies that they developed and used during this unit
3. Allow students to work at their own speed on the Posttest
4. Collect Posttests
5. At the end of the class pass back the team folders and allow each student to keep his/her Student Guide
6. Encourage them to keep the Student Guide and refer to it as needed

Speed Bumps

The speed bump marble-rolling competition involves rolling six marbles one meter, across your opponent's speed bump paper, and having the marbles land in a catching tray in which points are earned. The points are added to the points each team has earned while working on fraction speed bump challenges.

The speed bump marble-rolling competition brings a different kind of challenge to earning points. Teams may use their creativity and skill to stop opponents from gaining points by purchasing speed bumps and placing them strategically. Success with speed bumps may carry over to the Slow Marble Race, as teams remember what they did previously and use that information to win the culmination.

Information

1. Speed bumps make it more difficult for an opponent to roll a marble into the catching tray where it earns points
2. Teams spend points to buy speed bumps to thwart their opponent
3. There are different kinds of speed bumps
4. Teams may purchase speed bumps in any combination of heights
5. Teams need to purchase speed bumps and cut them out
6. Teams decide on the number of speed bumps and their placement, and then tape them down
7. Speed bumps are taped on an $8\frac{1}{2}$" x 11" inch sheet of paper (as determined by each team)
8. Speed bumps may be placed anywhere on the $8\frac{1}{2}$" x 11" paper, but may not touch each other
9. The $8\frac{1}{2}$" x 11" paper is taped between the starting line and the catching tray, with the first speed bump 50 cm from the front of the catching tray

Speed Bump Competition Options

You may decide to use any one of these repeatedly, or use them in succession as time allows.

Speed bumps allow teams to use physical skills and strategies to earn points by rolling marbles. You decide when to include a speed bump event in the schedule. There are different options for conducting a speed bump competition. Your decision may come down to the amount of time you have available.

Game overview for the first three options listed below:

1. No one may practice rolling the marbles across the speed bumps
2. Teams roll six marbles across or past the speed bump, trying to get any one of them to stop in the catching tray
3. Remove a marble if it stays in the tray, and record the fraction value
4. Once a marble lands on the largest fraction area, a team stops rolling
5. A team earns points based on the highest fraction area that any marble stops on. Multiply that fraction times the team's current point total and record those points on the team score sheet to get a new total.
6. Any marble that does not stop on the catching tray earns no points

First, Second, and Third Options (roll six marbles)—Premade Speed Bumps

1. **First option:** Tape one crooked speed bump to an $8\frac{1}{2}$" x 11" sheet of paper. Teams spend no points. They try to roll six marbles at the catching tray to try to earn points to add to their score sheet.
2. **Second option:** Have two crooked speed bumps premade and ready for teams to purchase. Teams purchase one speed bump for their opponents, trying to keep the marbles from reaching the catching tray. The easy speed bump costs 4, while the hard speed bump costs 8 points.
3. **Third option:** Have the three speed bumps (small, medium, and large) cut out and taped to three separate $8\frac{1}{2}$" x 11" sheets of paper before the competition begins
4. Teams select which premade speed bump (small, medium, or large) they want to purchase for their opponent's marble roll
5. Teams spend five points for a small speed bump, seven points for a medium speed bump, and three points for a large speed bump

Fourth Option—Team-Made Speed Bumps (roll six marbles)

Game overview:

1. The teams may buy up to two speed bumps (small, medium, or large) in any combination
2. A small bump cost five points, a medium bump costs seven points, and a large bump costs three points. Record the points spent on the team's score sheet.
3. A team may choose to spend no points and let their opponent roll at the catching tray without speed bumps in the way

Teaching tip
Many speed bump competitions take less than 15 minutes if the speed bumps are premade (that is, with 8 teams each rolling 6 marbles.)

Teaching tip
The roller (position rotates with each marble-rolling event: first the Recorder, then Setter, and then Checker) should release the marble from a six-inch starting line 100 cm from the front of the catching tray.

Teaching tip
To speed up the competition, make speed bump sheets ahead of time.

4. Once purchased, the team decides how it wants to place the speed bumps on an $8\frac{1}{2}$" x 11" sheet of paper; members then cut them out, fold them, and secretly tape them down. All teams must tape the bumps down at the same time, and no team may change its speed bumps. Any configuration is allowed, but the speed bump edges must be taped flat, with the minimum spacing (see speed bump patterns), and they may not not touch one another.

Teaching tip
Put a piece of tape 50 cm from the edge of the catching tray. Then it is easy to line up each speed bump sheet with the bump at the 50 cm mark.

5. The speed bump sheet is placed 50 cm from the front of the catching tray

6. The Roller should release the marble from a starting line 100 cm from the front of the catching tray

7. Mark the starting line with six inches of masking tape. All marbles should be released somewhere on that line of tape to prevent Rollers from taking an angle that always misses the speed bumps.

Teaching tip
Once teams have selected the speed bumps for their opponents, have all teams using the small speed bump go first. Then you do not have to change the speed bump paper until you get to teams that need the next-larger speed bump.

8. Decide how many "rolls" each team gets (we suggest six). Marbles that miss the tray earn nothing, but if a team lands more than one marble on the tray, it uses the largest fraction to compute its bonus points. If a marble lands on a line, the team gets the higher fraction. ***Option:*** *You could have three or four problems for the team to work correctly before it can record its newly earned points.*

9. Eight teams rolling six marbles takes less than 15 minutes

Fifth Option—Team-Made Speed Bump Trap (roll six marbles)

Game overview:

1. A speed bump trap may be used, instead of using only a speed bump sheet. The speed bump trap incorporates the speed bumps and catching tray into one unit.

2. Any number of speed bumps may be purchased for points, cut out, folded, and then taped down in the speed bump trap. A team must record on their score sheet the number of points spent on speed bumps.

3. Speed bumps may not touch when taped down in the speed bump trap

4. The team must divide the tray into four separate sections with approximately the same area in each

5. The teams write a different fraction in each area. Since the fraction becomes the multiplier for earning points, you must offer the four fractions ($\frac{1}{10}$, $\frac{1}{5}$, $\frac{1}{4}$, and $\frac{1}{6}$ are examples).

6. A team may skip using speed bumps and divide the trap into four areas with a fraction written in each area

7. If a marble lands on a line, the team gets the higher fraction as its multiplier for earning more points
8. The speed bump trap's front edge is taped down 100 cm from the starting line

Crooked Speed Bumps—Options 1 and 2

You might use a crooked speed bump sheet for options one and two. It consists of one crooked speed bump taped to an $8\frac{1}{2}$" x 11" sheet of paper.

Crooked speed bump examples:

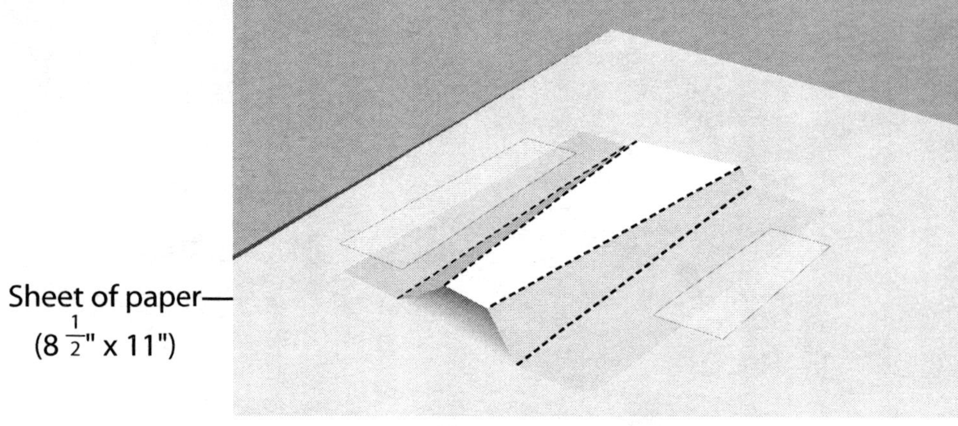

Hard crooked speed bump

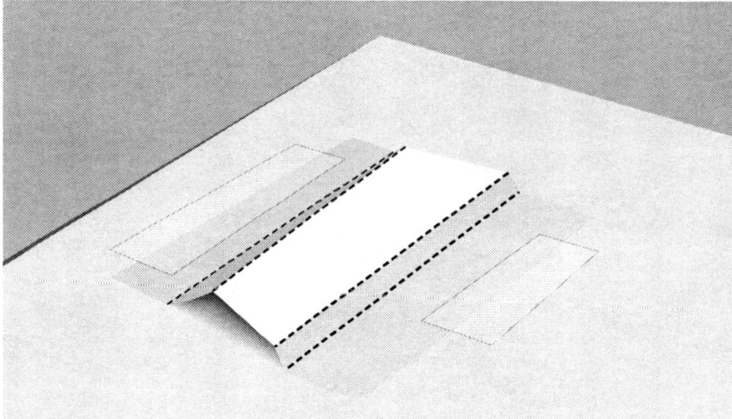

Easy crooked speed bump

Tape the two 1.5 cm x 9 cm rectangles flat on a sheet of paper. Fold on dashed lines to create the raised speed bump.

***Hard** crooked speed bump* (8 points):

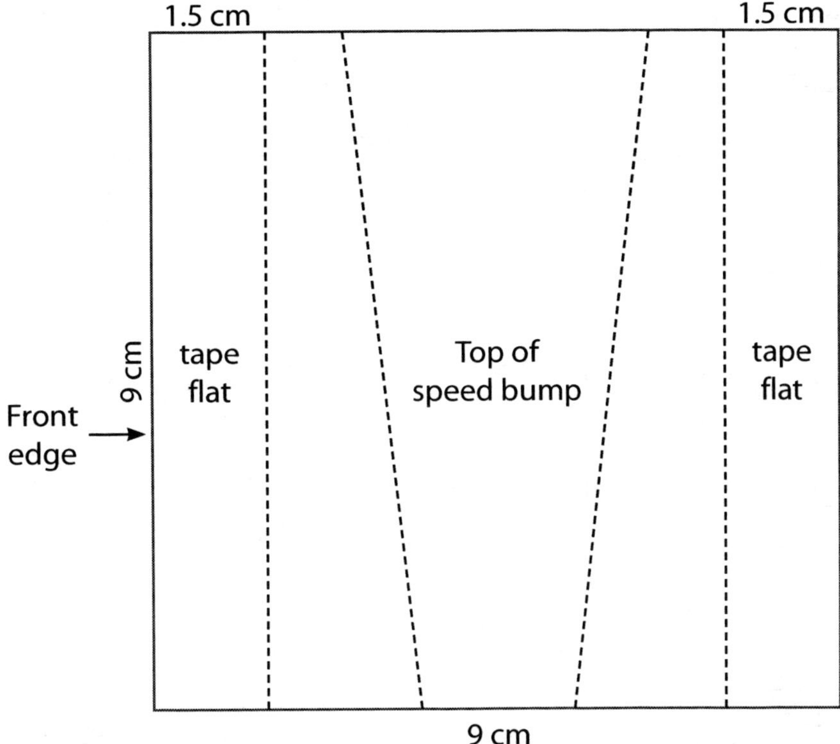

***Easy** crooked speed bump* (4 points)

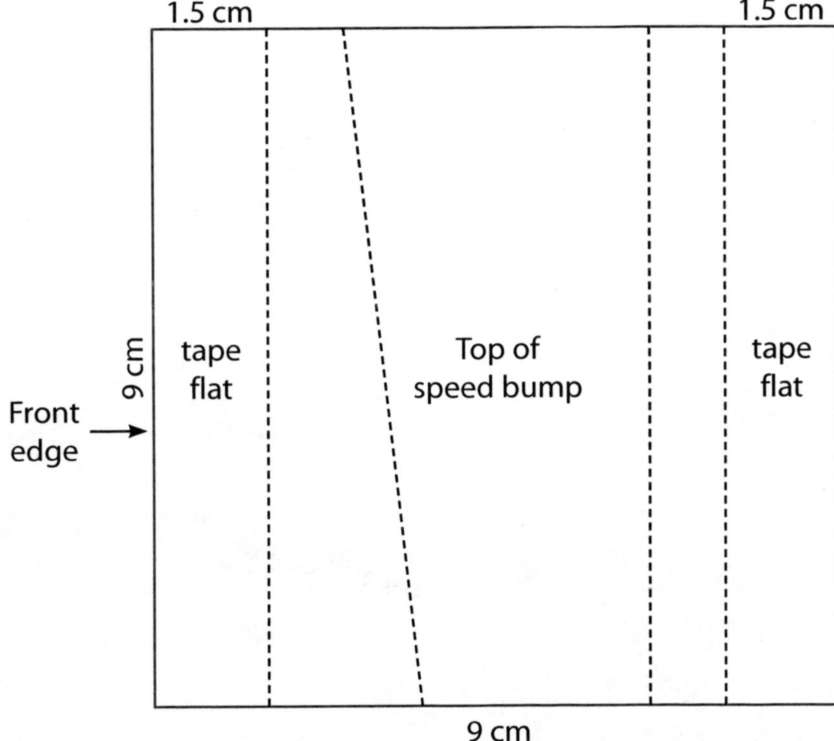

Speed Bump Patterns—Used for Options 3, 4, and 5

Fold on the dashed lines.

Small speed bump *(5 points)*

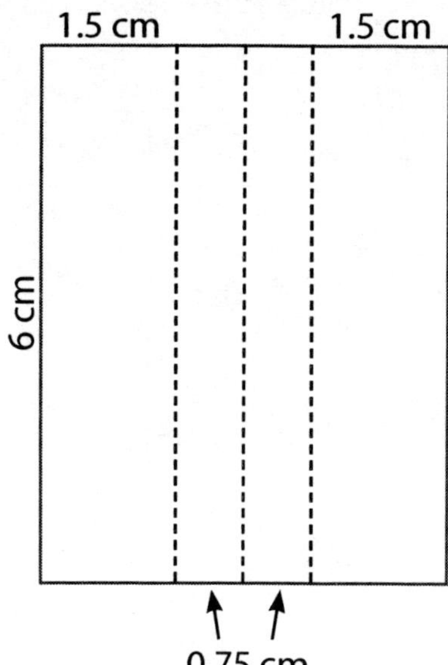

Medium speed bump *(7 points)*

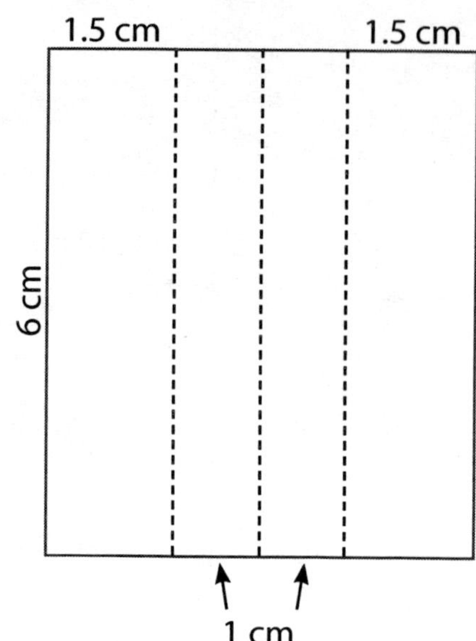

Teaching tip

Tell students: When folding the speed bumps, fold while looking at the lines and then fold back on the crease to get the desired shape. Do not forget to use a centimeter ruler and make sure that the speed bumps are taped down with the minimum spacing between the two edges.

Large speed bump *(3 points)*

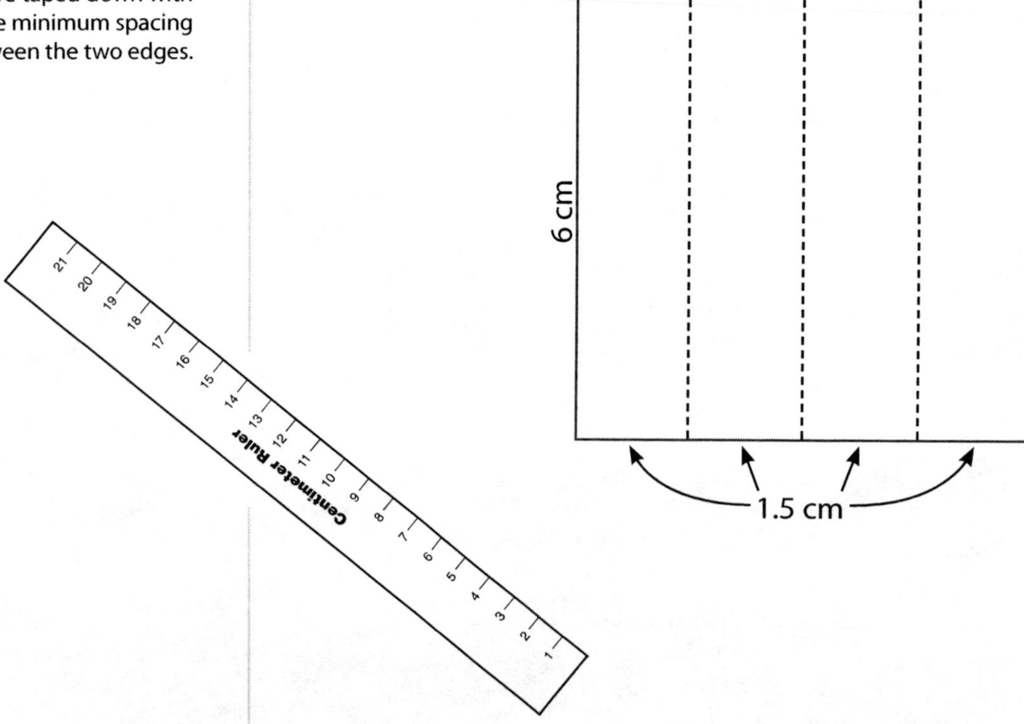

Important: When a speed bump is taped down, there must be a minimum distance between the two 6 cm sides. Students must measure and have opponents verify that the minimum spacing has been met to keep the competition fair. See below.

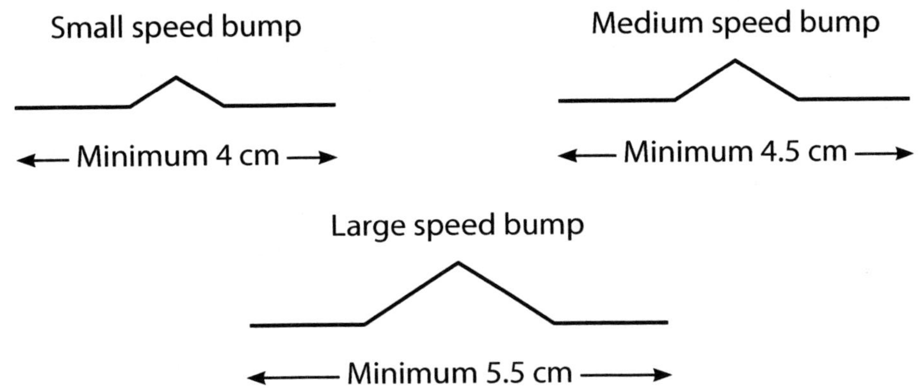

Catching Tray

Fold on dashed lines so there is a 1 cm edge (90 degrees) on three sides of the tray. Tape the corners.

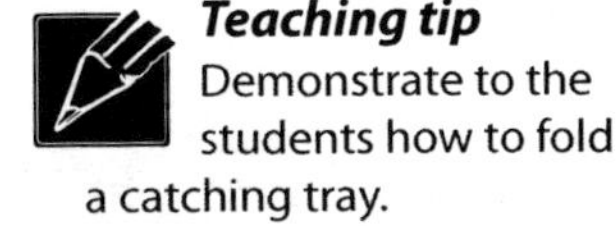

Teaching tip
Demonstrate to the students how to fold a catching tray.

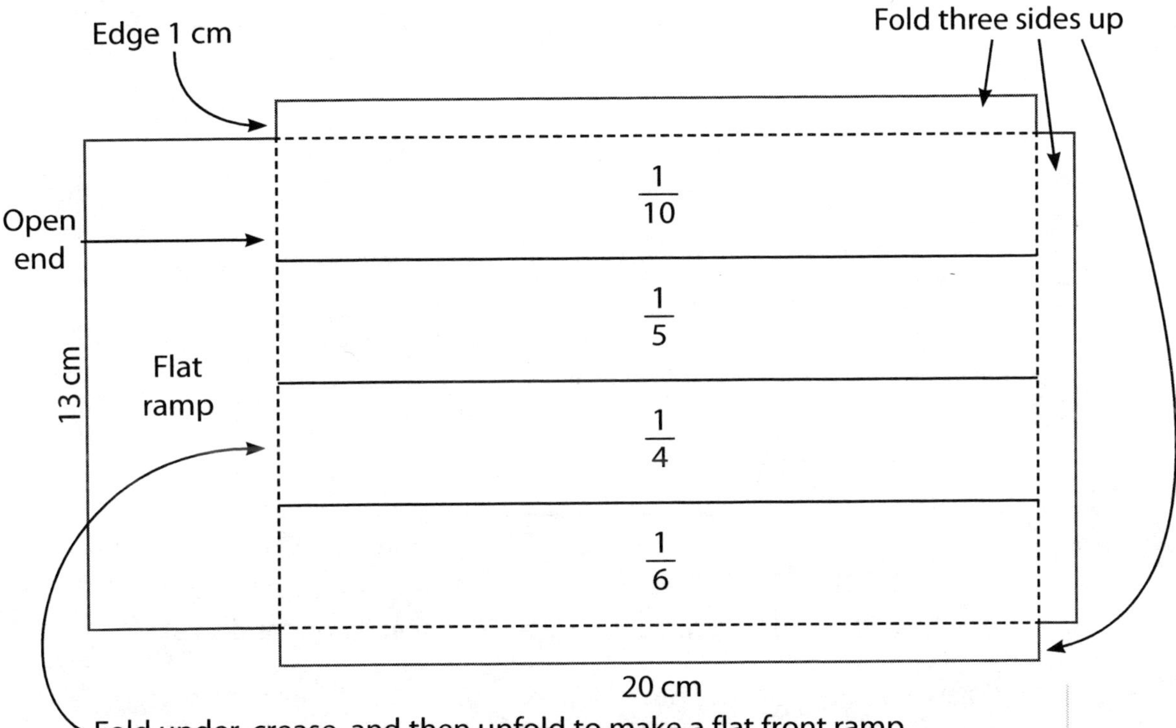

Fold the flat ramp under the box, crease the edge, and then unfold and tape the leading edge of the ramp down flat. This will make a slight rise at that front fold on the catching tray and keep some of the marbles from rolling off the tray. Even though the back edge of the tray is taped at the corners, some marbles will roll out the back end of the tray. Many marbles will hit the back edge and then rebound and roll off the tray.

Speed Bump Trap

Fold sides and tape corners

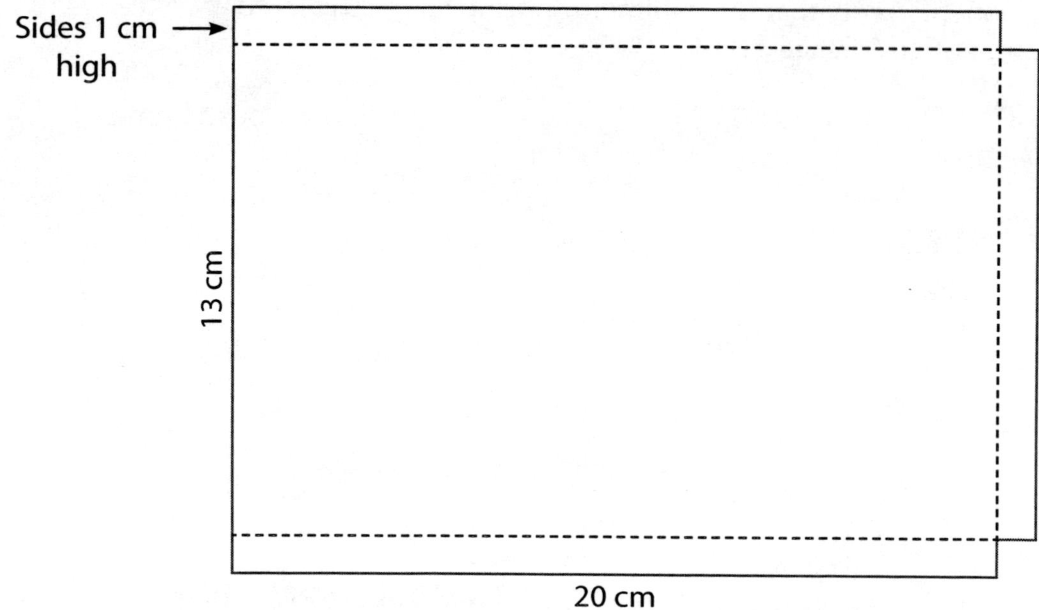

Speed Bump Trap Diagram (example):

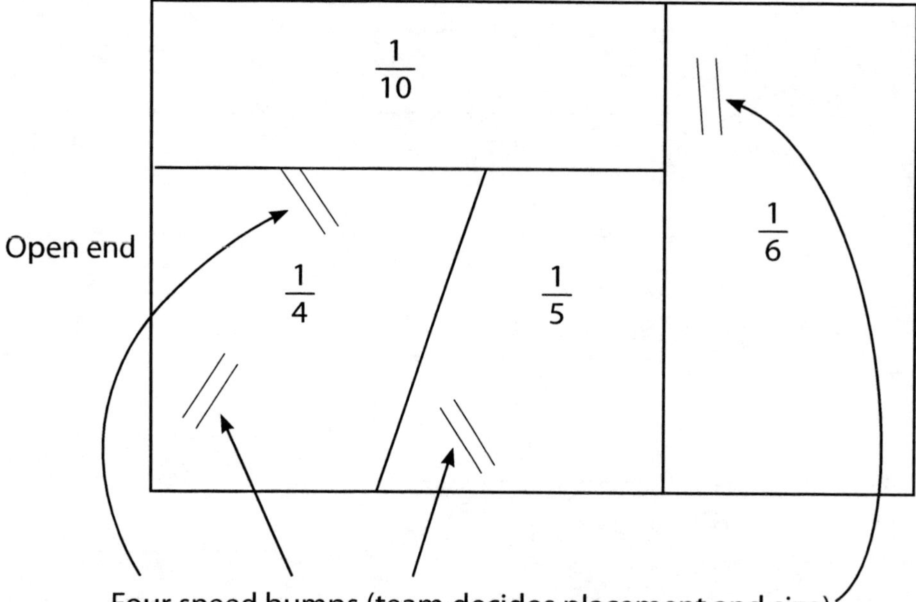

Each team decides how the paper will be divided into four sections (approximately the same area for each). Students draw the lines on the paper and write the fractions in the spaces before taping down the speed bumps.

Culmination
Slow Marble Race

The Slow Marble Race starts with your knowing the total points that each team has. You will determine the number of centimeters of track that each team will receive. Each team needs about 200 cm of track, but more will be acceptable. Example: If all of your teams have earned well over 200 points, then you need to determine how many points each cm of track will cost. 300 points divided by $1\frac{1}{2}$ equals 200. You can say that each cm of track costs $1\frac{1}{2}$ points. If your teams did not earn enough points, you can just give a bonus for teamwork to the whole class. Example: Teams earned about 135 points. Give a bonus of 75 to get them near 200. It is not necessary for teams to have the same number of points. A team with 167 points won over teams with 200 to 220 points.

You need to keep the amount of track material manageable. Students get 12" x 18" (or 18" x 24") sheets of heavy paper and make their track sections by drawing parallel lines. They then cut and fold all of the track material. They have the Checker from their opposing team verify the correct length they have earned from their points total. This material may be used in any way to create the slow marble track. The track must be folded to the normal "U" shape, measuring 2 cm x 2 cm x 2 cm. (See Student Guide.) Part of the track will be used as support pillars for keeping the track elevated and stable. Some of the track may be cut up and then used (in any way) to slow down the marble as it makes its descent. Any method to slow the marble is acceptable. Teams may only use clear tape and the track material to build the racetrack.

Once a team's points are converted to track, the team may not get any more paper.

Teams may cut slits in the track to create a curve, a turn, or a dip in the track. (See Student Guide.)

Races

Races may be run continually, because teams are really racing against the clock to get a slower time. Use a stopwatch to record each team's best (slowest) marble run. During the culmination, you can record each team's improvement on the board next to their team name. With modifications to their track design, most teams will improve their best times.

This method of test, revise, and retest keeps each team trying to better their own best time as well as move up in the class standings. More teams will be involved until the last minute.

Culmination Option

If you want to focus mainly on fractions and not attempt the Slow Marble Race, you may still run a culmination involving rolling marbles over two speed bumps into a catching tray. This event involves all teams building a catching tray and a speed bump sheet. You use all of the regular rules from earlier in the unit. Their opponents get to have all team members roll four marbles (12 per team). Each team uses the fractions earned (by landing in the catching tray) to create the largest final answer by using addition, subtraction, multiplication, and/or division. Use each fraction once. As in the challenge sheets, two fractions are used with any symbol to create an answer. Teams take that new answer plus another remaining fraction and create another larger answer until all fractions are used once. Concepts learned during the speed bump challenges may help a team that has fewer fractions to generate the largest final answer. The team with the greatest answer wins.

Answer Key
Pretest/Posttest

Teaching tip
Urge students to skip the ones they do not know and work on the ones they feel comfortable with.

For numbers 1 through 5, circle the answers you choose:

1. Which two fractions are the largest?

   $\frac{9}{12}$ (circled) $\frac{7}{10}$ $\frac{2}{20}$ $\frac{3}{13}$ $\frac{5}{7}$ $\frac{4}{8}$ $\frac{4}{5}$ (circled) $\frac{4}{7}$

2. Which two fractions, when added, will give the smallest answer?

   $\frac{2}{13}$ $\frac{1}{11}$ (circled) $\frac{4}{40}$ (circled)

3. Select two fractions that will give the largest answer when subtracted.

   $\frac{1}{7}$ (circled) $\frac{3}{8}$ $\frac{4}{5}$ $\frac{9}{10}$ (circled)

4. Which two of these can be multiplied to get an answer less than $\frac{1}{2}$?

   $\frac{7}{8}$ $\frac{3}{5}$ (circled) $\frac{2}{3}$ (circled) $\frac{19}{20}$

5. How many of these can you use as divisors to get answers less than $\frac{1}{4}$?

   $\frac{1}{3}$ divided by… $\frac{1}{2}$ $\frac{1}{4}$ 2 (circled)

6. Reduce these fractions and explain how you reduced them:

   a. $\frac{16}{32}$ $\frac{1}{2}$ …16 is half of 32

   b. $\frac{40}{90}$ $\frac{4}{9}$ …divide by ten (cross out the zeros)

   c. $\frac{7}{21}$ $\frac{1}{3}$ …divide by one $\left(\frac{7}{7}\right)$

7. Fill in the missing values:

   $$\frac{2}{\mathbf{5}} = \frac{\mathbf{6}}{15} = \frac{8}{\mathbf{20}} = \frac{\mathbf{200}}{\mathbf{500}}$$

8. Complete this adding problem:

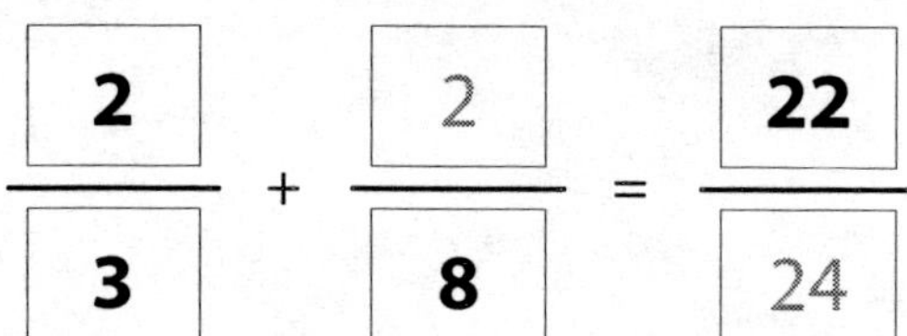

9. Use these numbers: (4) 5 12 (16)

Select two and write the smallest fraction.

$$\frac{4}{16}$$

10. Use these numbers: 12 15 (19) (21)

Select two and write the largest fraction.

$$\frac{19}{21}$$

11. Complete this subtraction problem:

$$\begin{array}{r} 7\frac{\boxed{35}}{70} \\ -\;3\frac{8}{16} \\ \hline 4 \end{array}$$

Answer Key

Teamwork Practice: Day 2

Use regular calculators. Focus on teamwork.

Place the values in the boxes to create the highest final answer. The answer to each problem goes in the circle and slides to the next circle to begin another math problem. All four symbols (+ – × ÷) go in the diamonds. Your team's best final answer will be added to your team's score sheet.

1. The class must use these five values: 6, 7, 8, 9, 10

EXAMPLE:

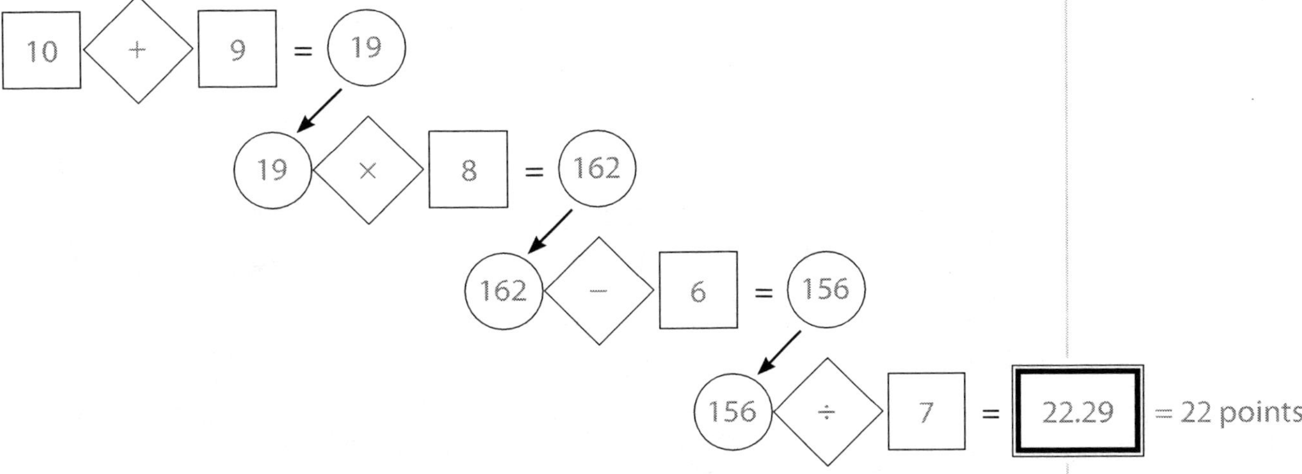

2. This time no symbol may be placed in the same diamond position as in the first sequence. Work for the highest answer. Your team's best final answer will be added to your team's score sheet.

3. This time the opposing team's Setter will place two different symbols in two of your diamonds before your team may begin. Your team then places the other two symbols and all four values in the boxes. Work for the highest answer. Your team's best final answer will be added to your team's score sheet.

Answer Key
Fractions Speed Bump: Day 3

Use the values provided and fill in the boxes to make the problem correct. Each value may be used only once. Earn the most points for being the first team to finish correctly. Teams all earn points but gain fewer points than those who have already finished correctly.

Values: 1, 2, 4, 5, 6, 8

$$\frac{\mathbf{4}}{8} = \frac{1}{\mathbf{2}} = \frac{2}{4} = \frac{\mathbf{3}}{6} = \frac{5}{\mathbf{10}}$$

Values: 2, 4, 6, 8, 10, 12, 14

$$\frac{2}{6} = \frac{8}{\mathbf{24}} = \frac{10}{\mathbf{30}} = \frac{12}{\mathbf{36}} = \frac{14}{\mathbf{42}} = \frac{4}{\mathbf{12}}$$

Values: 1, 1, 2, 5, 11, 20

$$\frac{1}{5} + \frac{2}{\mathbf{8}} + \frac{1}{\mathbf{10}} = \frac{\mathbf{22}}{\mathbf{40}} = \frac{11}{20}$$

Answer Key
Adding Fractions Speed Bump: Day 4

Use each value once. Place the values in the boxes so that you create a correct mathematical sentence.

1. **Values: 1, 2, 3, 4, 5, 8**

$$\frac{1}{\mathbf{3}} + \frac{2}{4} + \frac{5}{\mathbf{6}} = \frac{\mathbf{20}}{\mathbf{12}} = \mathbf{1}\frac{8}{\mathbf{12}} = \mathbf{1}\frac{\mathbf{2}}{3}$$

2. **Values: 1, 2, 4, 5, 6, 12**

$$\frac{2}{4} + \frac{1}{5} + \frac{\mathbf{3}}{6} = \frac{\mathbf{72}}{\mathbf{60}} = \mathbf{1}\frac{12}{\mathbf{60}} = \mathbf{1}\frac{\mathbf{1}}{\mathbf{5}}$$

3. Use these values to find the largest possible answer. Use proper fractions only. You may not use whole numbers ($\frac{3}{1}$).

Values: 1, 2, 4, 5, 6, 12, 25

$$\frac{\mathbf{3}}{4} + \frac{5}{6} + \frac{1}{2} = \frac{25}{12} = \mathbf{2}\frac{\mathbf{1}}{\mathbf{12}}$$

4. Use the values to find the largest answer and the smallest answer. Use proper fractions only. You may not use any whole-number designations (such as $\frac{3}{1}$). Discuss a team strategy before beginning. Points are awarded for teams that finish quickly. Additional points are awarded if your team has the correct largest and/or smallest answer.

Values: 2, 3, 4, 5, 6, 7

$$\frac{6}{7} + \frac{4}{5} + \frac{2}{3} = \frac{244}{105} \quad \text{largest}$$

$$\frac{2}{5} + \frac{3}{6} + \frac{4}{7} = \frac{309}{210} \quad \text{smallest}$$

Answer Key

Subtracting Fractions: Day 6

1. Use only four of the eight values given. Create two proper fractions that when subtracted will give the **largest** answer available. If finished early, go for the bonus assignment. ***Bonus:*** Use the other four values (not yet used in the first challenge) and find the smallest answer you can. Team(s) with the **smallest** answer will gain bonus points to add to their total.

Values: 2, 4, 5, 6, 8, 10, 15, 20

$$\frac{5}{6} - \frac{2}{20} = \boxed{\frac{44}{60}}$$

Reduces to $\frac{22}{30}$ then $\frac{11}{15}$

Bonus:

$$\frac{10}{15} - \frac{4}{8} = \boxed{\frac{20}{120}}$$

Reduces to $\frac{2}{12}$ then $\frac{1}{6}$

2. Use all of the values once. Use four of the values to create two proper fractions to use in the first subtraction problem. After solving the problem, use that answer and subtract another proper fraction by using the remaining two values. Use experimentation to find the **smallest** answer available. No negative answers allowed.

Values: 1, 2, 3, 4, 5, 6

$$\frac{5}{6} - \frac{1}{3} = \frac{3}{6}$$

$$\frac{3}{6} - \frac{2}{4} = \boxed{0}$$

3. Subtract then add. This is a challenge to build the **largest** answer possible. Use all of the values once. Use four of the values to create two proper fractions to use in the subtraction problem. Use the two remaining values to make a proper fraction that ***when added*** to the first answer will give the **largest** final answer.

Values: 1, 2, 4, 8, 10, 12

$$\frac{2}{4} - \frac{1}{12} = \frac{5}{12}$$

$$\frac{5}{12} + \frac{8}{10} = \boxed{\frac{73}{60}}$$ largest

Teaching tip
There may be more than one way to use the values to generate a specific high or low answer.

Answer Key

Multiplying Fractions Speed Bump: Day 7

1. Use the values and experiment with multiplying proper fractions. Search for the largest and smallest answer. After working on the challenge, your team will need to decide if it wants to commit to having the largest or smallest answer. You must circle the word "largest" or "smallest" on your paper. You gain points if your selected answer is the largest or the smallest. If only a few teams match the correct answer (low and high) the points earned will be larger. Many correct answers to "largest" or "smallest" will reduce the points earned. Good luck choosing and working correctly.

Values: 2, 4, 6, 8, 10, 12

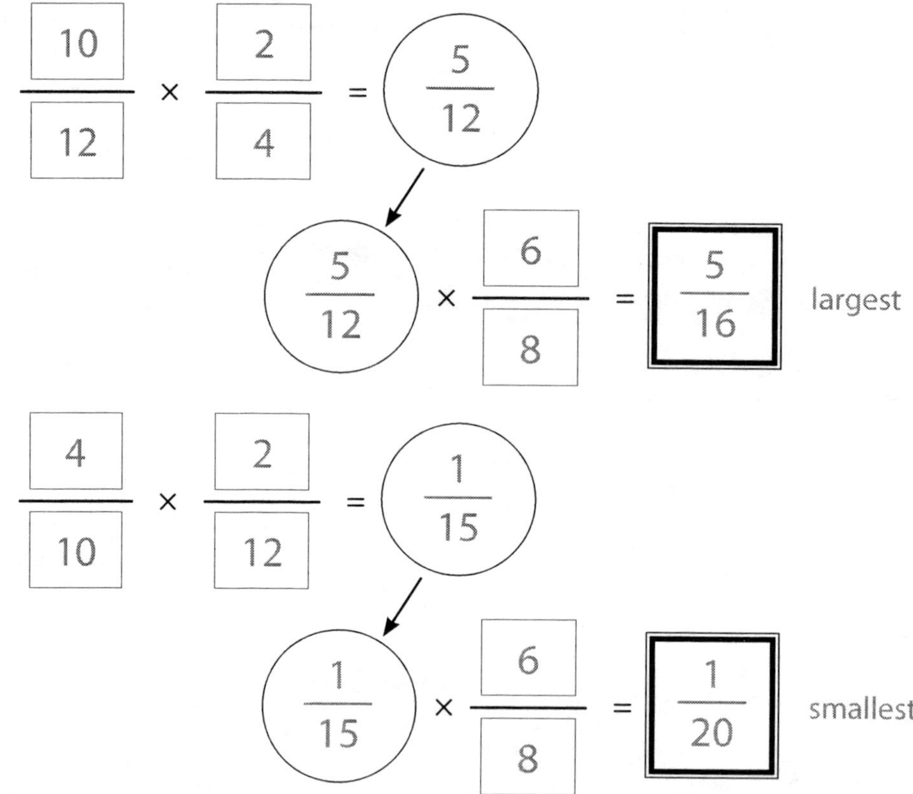

Teaching tip
Allow time for team discussion before starting challenge two.

2. **Second challenge:** Use the same six values as provided above. First your opponent's Setter places two different *values* in any two of your team's fraction squares and they may not be moved. Your team must use two different values to try to generate a **large** answer. The larger answers score more points. **Proper fractions only.**

Answers will vary. You will need to join the class in comparing answers and determining the highest.

Teaching tip
Allow time for team discussion before starting challenge three.

3. **Third challenge:** Continue using these six values (2, 4, 6, 8, 10, 12). This time, your opponent's Setter places four of the given *values* in any of the six squares in the problem below. Your team may choose to replace one of the "placed" values with either of the two remaining values. You do not have to change a value. Again, try to generate a **large** answer. If your answer is greater than your opponent's answer, you win. You also win if you have the largest answer in the class. **Improper fractions allowed** for this challenge.

   Because of all of the possibilities, you and the class will have to determine the winning answer.

   Note: Information for the Setters for the second challenge. *The "12" may never be placed in a numerator box. If the "12" is in a denominator box, the "10" may never be placed in the opposite side (diagonal) numerator box. If this happens, you have violated the "proper fractions only" parameters.*

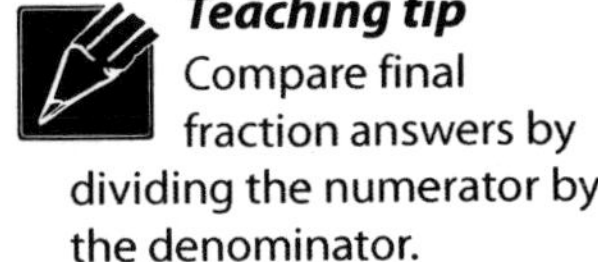

Teaching tip
Compare final fraction answers by dividing the numerator by the denominator.

Answer Key

Dividing Fractions Speed Bump: Day 9

1. Use the four values to create the division problem that gives the **largest** answer for the two challenges below. When you are finished try the bonus challenge. For the bonus, points will be awarded for the answer closest to $1\frac{7}{16}$.

Values: 2, 4, 6, 8—To be used with the three challenges below.

Proper fractions only:

$$\frac{4}{6} \div \frac{2}{8} = 2\frac{2}{3}$$

Improper fractions allowed:

$$\frac{8}{2} \div \frac{4}{6} = 6$$

Also $\frac{6}{2} \div \frac{4}{8} = 6$ or $\frac{6}{4} \div \frac{2}{8} = 6$

Bonus:

$$\frac{6}{8} \div \frac{2}{4} = 1\frac{1}{2}$$

Also $\frac{4}{8} \div \frac{2}{6} = 1\frac{1}{2}$ or $\frac{4}{2} \div \frac{8}{6}$

Teaching tip
Allow time for team discussion before starting challenge two.

2. Use the six values to create the **smallest** answer. If you believe you have the smallest answer, you may try for the bonus. If you do have the smallest answer, then you win bonus points, but if another team has a smaller answer, then you lose some points. Circle the word "smallest" below if you want to compete for the bonus points.

   Values: 1, 3, 5, 7, 8, 10

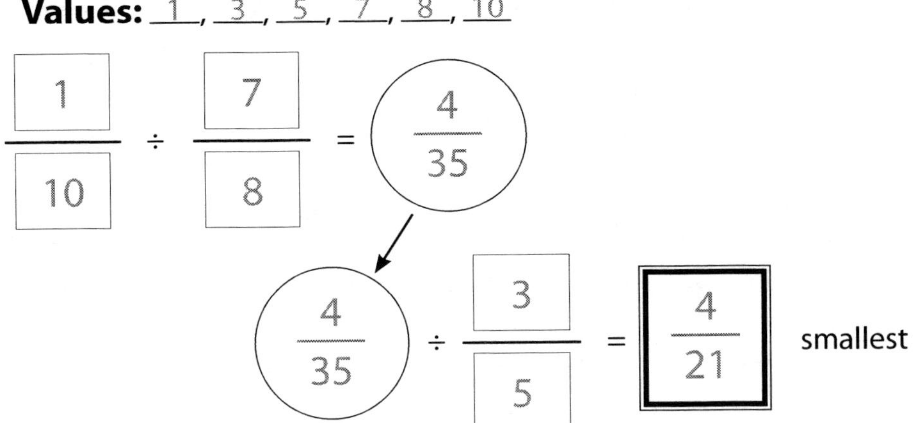

Teaching tip
Allow time for team discussion before starting challenge three

3. Try for **the largest or the smallest** answer this time. Use the six values given to create your answer. Your opponent's Setter will decide the placement of the two different symbols (+ – × ÷, placed in the diamonds) and you may not move them. All teams that find either the largest or the smallest answer will gain points. All ties earn points.

   Values: 3, 5, 7, 8, 9, 10

   Because of all of the possibilities, you and the class will have to determine the winning answer.

4. Second sequence: Your opponent's Setter must pick a different combination of two different symbols (+ – × ÷). Use the same values again. All teams that find either the largest or the smallest answer will gain points. All ties earn points.

   Because of all of the possibilities, you and the class will have to determine the winning answer.

Answer Key

Addition Extension Activity 1(e and h)

Problems are listed from easy to hard.

Depending on the class, the teacher may want use either sheet 1e (easier) or sheet 1h* (harder—because the fraction answers have been reduced making it more difficult to solve). Whole numbers and fractions may be in either addend spot. **Option:** The teacher may require the students to write the equivalent fractions next to the fraction boxes when they figure out the common denominators.

Whole numbers: 2, 5, 7, 8 **Fractions: $\frac{1}{3}$, $\frac{2}{8}$, $\frac{3}{4}$, $\frac{4}{9}$**

You may use any fraction or any whole number twice in any problem.

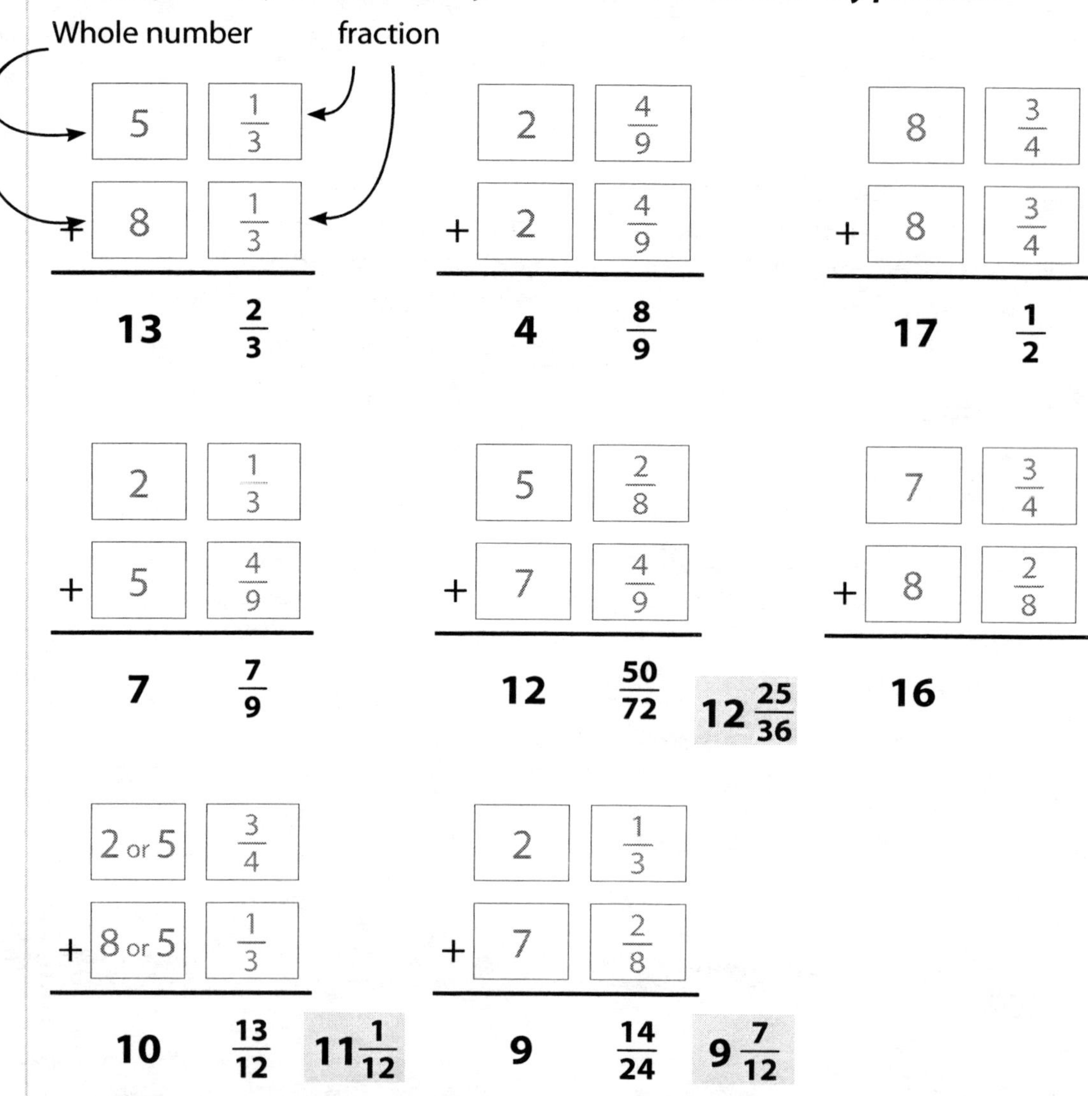

Fractions in gray boxes are the answers in reduced form and they match the Addition Extension sheet 1h answers.

Answer Key
Subtraction Extension Activity 1 (e and h)

Problems are listed from easy to hard.

Depending on the class, the teacher may want use either sheet 1e (easier) or sheet 1h (harder—because the fraction answers have been reduced making it more difficult to solve). **Option:** The teacher may require the students to write the equivalent fractions next to the fraction boxes when they figure out the common denominators.

Whole numbers: 3, 4, 6, 9 Fractions: $\frac{2}{10}$, $\frac{4}{6}$, $\frac{9}{12}$, $\frac{5}{10}$

You may not use any fraction or any whole number twice in any one problem.

Whole number fraction

$4\ \frac{5}{10} - 3\ \frac{2}{10} = 1\ \frac{3}{10}$

$6\ \frac{9}{12} - 4\ \frac{4}{6} = 2\ \frac{1}{12}$

$6\ \frac{4}{6} - 3\ \frac{2}{10} = 3\ \frac{14}{30}$ — $3\ \frac{7}{15}$

$6\ \frac{9}{12} - 3\ \frac{5}{10} = 3\ \frac{15}{60},\ 3\ \frac{3}{12}$ — $3\ \frac{1}{4}$

$9\ \frac{9}{12} - 4\ \frac{2}{10} = 5\ \frac{33}{60}$ — $5\ \frac{11}{20}$

$9\ \frac{2}{10} - 3\ \frac{5}{10} = 5\ \frac{7}{10}$

regrouping

$9\ \frac{4}{6} - 6\ \frac{9}{12} = 2\ \frac{11}{12}$

regrouping

$9\ \frac{2}{10} - 4\ \frac{4}{6} = 4\ \frac{16}{30}$ — $4\ \frac{8}{15}$

regrouping

Fractions in gray boxes are the answers in reduced form and they match the Subtraction Extension sheet 1h answers.

Answer Key

Addition and Subtraction Extension Activity

Add and subtract only

Use the fractions provided to generate answers that are on the game board. Any three in a row wins points, but the more answers you find, the higher your point total. On this paper write a math problem that gives an answer that matches a fraction on the game board. Put an "x" over the matching fraction on the game board. Then write a number "1" by your first answer and by the "x"ed-out answer on the game board. Continue numbering your fraction problems and answers that match using this method. Fractions can be used repeatedly in any combination.

Fractions: $\frac{1}{2}, \frac{1}{3}, \frac{1}{4}, \frac{1}{5}, \frac{1}{6}, \frac{1}{7}, \frac{1}{8}$

$\frac{1}{6}$	$\frac{7}{10}$	$\frac{11}{24}$
$\frac{12}{35}$	$\frac{5}{6}$	$\frac{1}{12}$
$\frac{7}{12}$	$\frac{5}{12}$	$\frac{3}{8}$

Row one

$\frac{1}{3} + \frac{1}{4} = \frac{2}{12} = \frac{1}{6}$

$\frac{1}{2} + \frac{1}{5} = \frac{7}{10}$

$\frac{1}{8} + \frac{1}{3} = \frac{11}{24}$

Row two

$\frac{1}{5} + \frac{1}{7} = \frac{12}{35}$

$\frac{1}{2} + \frac{1}{3} = \frac{5}{6}$

$\frac{1}{4} - \frac{1}{6} = \frac{1}{12}$ or $\frac{1}{3} - \frac{1}{4} = \frac{1}{12}$

Row three

$\frac{1}{6} + \frac{1}{8} = \frac{14}{24} = \frac{7}{12}$

$\frac{1}{4} + \frac{1}{6} = \frac{5}{12}$

$\frac{2}{4} - \frac{1}{8} = \frac{3}{8}$

There are other answers besides the ones listed above.

Answer Key
Multiplication Extension Activity

Use the values given and fill in the boxes so that you create a correct math problem.

Each number listed is used once. Proper fractions only.

3, 3, 4, 5, 9, 20

$$\frac{3}{4} \times \frac{3}{5} = \frac{9}{20}$$ or $\frac{3}{5} \times \frac{3}{4}$ is acceptable

1, 2, 3, 4, 4, 6, 12, 24

$$\frac{4}{6} \times \frac{3}{4} = \frac{12}{24} = \frac{1}{2}$$ or $\frac{3}{4} \times \frac{4}{6}$ is acceptable

2, 3, 5, 6, 8, 10, 40, 60

$$\frac{8}{10} \times \frac{5}{6} = \frac{40}{60} = \frac{2}{3}$$ or $\frac{5}{6} \times \frac{8}{10}$ is acceptable

5, 7, 7, 30, 49, 6

$$\frac{6}{7} \times \frac{5}{7} = \frac{30}{49}$$ or $\frac{5}{7} \times \frac{6}{7}$ is acceptable

Answer Key
Multiplying With Fractions Extension

If your multiplier is a proper fraction, your answer will be smaller than the number you started with. When you multiply using a fraction you get PART of your original value. Your answer is less than your original value.

Start with16 and multiply by $\frac{1}{2}$.

You are really asking, "What is half of sixteen?"

Original value × the fraction $\frac{1}{2}$ = answer

16	$\frac{1}{2}$	8
8	$\frac{1}{2}$	4
4	$\frac{1}{2}$	2
2	$\frac{1}{2}$	1
1	$\frac{1}{2}$	$\frac{1}{2}$
$\frac{1}{2}$	$\frac{1}{2}$	$\frac{1}{4}$
$\frac{1}{4}$	$\frac{1}{2}$	$\frac{1}{8}$
$\frac{1}{8}$	$\frac{1}{2}$	$\frac{1}{16}$
$\frac{1}{16}$	$\frac{1}{2}$	$\frac{1}{32}$

Notice that multiplying by $\frac{1}{2}$ gives the same answer as dividing by 2. See Student Guide, page 11.

Use the information from the table above to complete the tables below:

16	$\frac{1}{16}$	1
8	$\frac{1}{16}$	$\frac{1}{2}$
4	$\frac{1}{16}$	$\mathbf{\frac{1}{4}}$
2	$\frac{1}{16}$	$\frac{1}{8}$
1	$\frac{1}{16}$	$\frac{1}{16}$
$\mathbf{\frac{1}{2}}$	$\frac{1}{16}$	$\frac{1}{32}$
$\mathbf{\frac{1}{4}}$	$\frac{1}{16}$	$\frac{1}{64}$
$\mathbf{\frac{1}{8}}$	$\frac{1}{16}$	$\frac{1}{128}$
$\mathbf{\frac{1}{16}}$	$\frac{1}{16}$	$\frac{1}{256}$

$\frac{8}{3}$	$\frac{1}{6}$	$\frac{8}{18}\left(\frac{4}{9}\right)$
$\frac{4}{3}$	$\frac{1}{6}$	$\frac{4}{18}\left(\frac{2}{9}\right)$
$\mathbf{\frac{2}{3}}$	$\frac{1}{6}$	$\frac{2}{18}\left(\frac{1}{9}\right)$
$\frac{1}{3}$	$\frac{1}{6}$	$\frac{1}{18}$
$\frac{1}{6}$	$\frac{1}{6}$	$\frac{1}{36}$
$\frac{1}{12}$	$\frac{1}{6}$	$\frac{1}{72}$
$\frac{1}{24}$	$\frac{1}{6}$	$\mathbf{\frac{1}{144}}$

Answer Key
Division Extension

1, 2, 4, 12, 24

$$\frac{12}{1} \div \frac{2}{4} = 24$$

1, 3, 3, 8, 8

$$\frac{3}{8} \div \frac{1}{8} = 3$$

2, 3, 8, 10—These are the values for the division problem only. You must provide the other numbers as you figure out the problem.

$$\frac{8}{10} \div \frac{2}{3} = \frac{12}{10} = 1\frac{2}{10} = 1\frac{1}{5}$$

(also $\frac{24}{20}$) (also $\frac{4}{20}$)

1, 1, 1, 4, 8, 32

$$\frac{4}{1} \div \frac{1}{8} = \frac{32}{1}$$

Answer Key
Dividing With Fractions Extension

When dividing by a fraction, your answer will be larger if the original value (dividend) is a whole number. When dividing a fraction by a smaller fraction, your answer will be larger than the dividend. When dividing a fraction by a larger fraction, the answer will be larger than the dividend.

Start with 10 and divide by $\frac{1}{2}$.
You are really asking, "How many one-halves are in ten?" The answer is 20.

Start with $\frac{1}{2}$ and divide by $\frac{1}{4}$.
You are really asking, "How many one-fourths are in one half?" The answer is two.

Start with $\frac{1}{4}$ and divide by $\frac{1}{2}$.
You are really asking, "How many one-halves are in one fourth?" The answer is one-half. It takes one-half of a half to make one fourth.

Original value ÷	the fraction $\frac{1}{2}$ =	answer
16	$\frac{1}{2}$	32
8	$\frac{1}{2}$	16
4	$\frac{1}{2}$	8
2	$\frac{1}{2}$	4
1	$\frac{1}{2}$	2
$\frac{1}{2}$	$\frac{1}{2}$	1
$\frac{1}{4}$	$\frac{1}{2}$	$\frac{1}{2}$
$\frac{1}{8}$	$\frac{1}{2}$	$\frac{1}{4}$
$\frac{1}{16}$	$\frac{1}{2}$	$\frac{1}{8}$

When dividing, you are always asking, "How many of this fraction does it take to make this value?"

Use the information from the table above and complete the tables below:

16	$\frac{1}{8}$	128
8	$\frac{1}{8}$	64
4	$\frac{1}{8}$	32
2	$\frac{1}{8}$	**16**
1	$\frac{1}{8}$	8
$\mathbf{\frac{1}{2}}$	$\frac{1}{8}$	4
$\mathbf{\frac{1}{4}}$	$\frac{1}{8}$	2
$\mathbf{\frac{1}{8}}$	$\frac{1}{8}$	1
$\mathbf{\frac{1}{16}}$	$\frac{1}{8}$	$\frac{1}{2}$

24	4	6
12	4	3
6	4	$1\frac{1}{2}$
3	4	$\mathbf{\frac{3}{4}}$
$1\frac{1}{2}$	4	$\frac{3}{8}$
$\frac{3}{4}$	4	$\frac{3}{16}$
$\frac{3}{8}$	4	$\frac{3}{32}$

Answer Key
Fraction to Decimal Practice Extension

Try to match each fraction with its decimal equivalent and then check your work using a calculator. Draw lines from each fraction to the decimal equivalent.

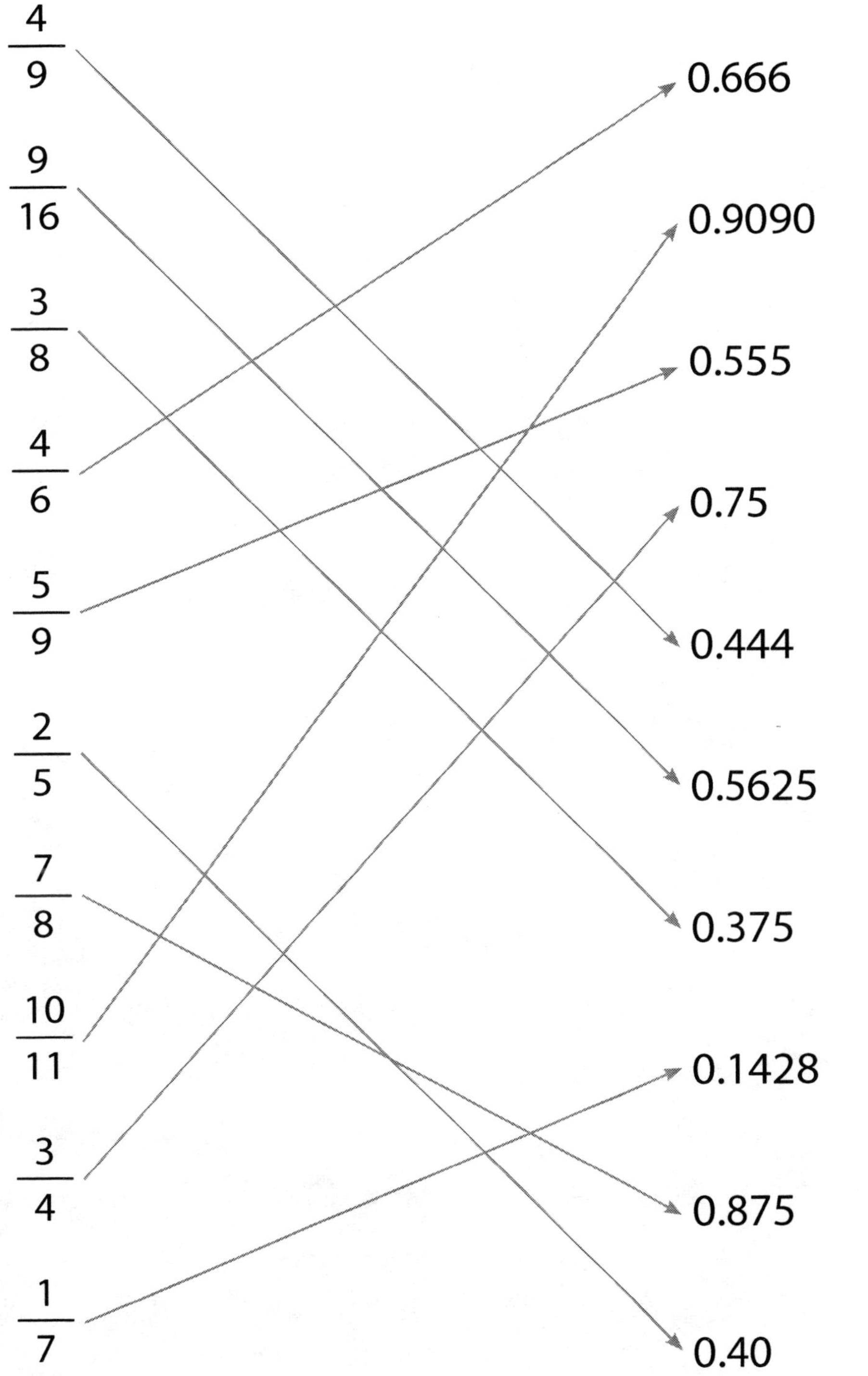

Answer Key

Fraction to Decimal Extension

Use the values given to fill in the numerator box in each fraction so that you create a fraction that equivalent to one of the decimal answers in the right-hand column. Draw a line from the newly created fraction to the proper decimal. Use each value once.

Values: 3, 9, 2, 10, 7, 4

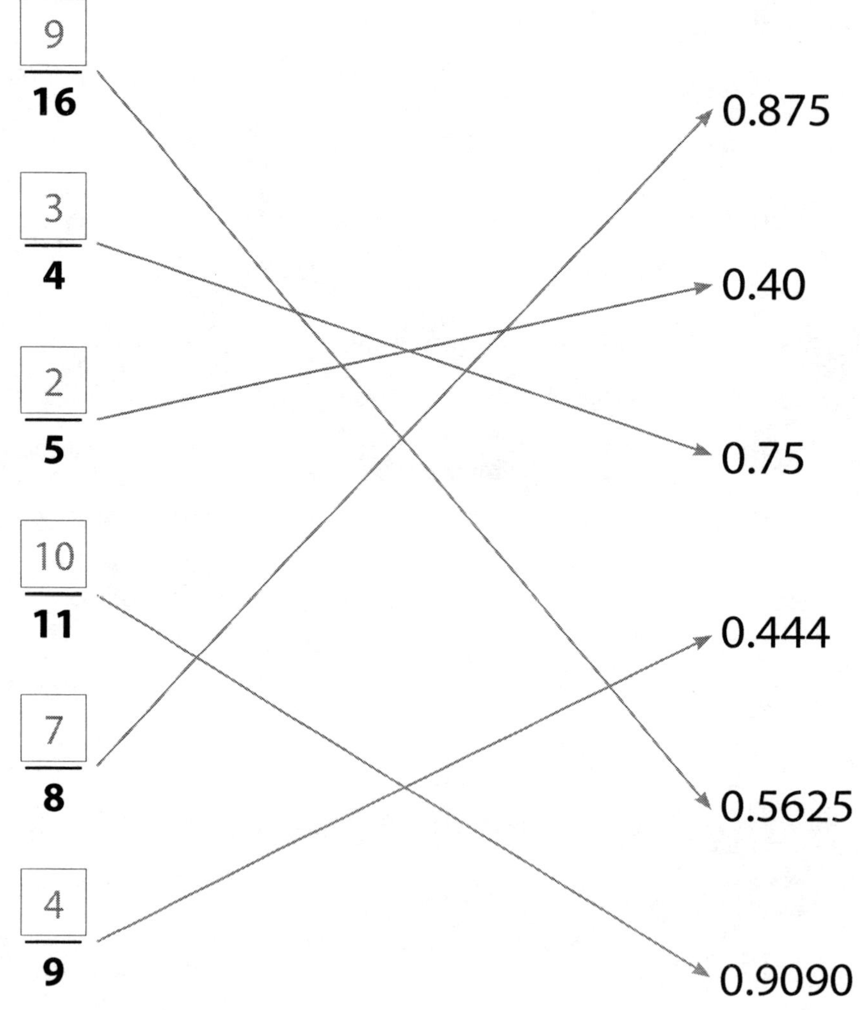

Answer Key

Multiplication and Division Extension

Multiply and divide only

Use the fractions provided to generate answers that are on the game board. Any three in a row wins points, but the more answers you find, the higher your point total. On this paper, write a math problem that gives an answer that matches a fraction on the game board. Put an "x" over the matching fraction on the game board. Then write a number "1" by your first answer and by the "x"ed-out answer on the game board. Continue numbering your fraction problems and answers that match using this method. Fractions can be used repeatedly in any combination.

Fractions: $\frac{1}{2}, \frac{2}{3}, \frac{3}{4}, \frac{4}{5}, \frac{5}{6}, \frac{6}{7}, \frac{7}{8}$

$\frac{4}{3}$	$\frac{2}{6}$	$\frac{15}{24}$
$\frac{24}{35}$	$\frac{4}{7}$	$\frac{21}{16}$
$\frac{7}{16}$	$\frac{14}{15}$	$\frac{20}{21}$

$\frac{2}{3} \div \frac{1}{2} = \frac{4}{3}$

$\frac{1}{2} \times \frac{2}{3} = \frac{2}{6}$

$\frac{3}{4} \times \frac{5}{6} = \frac{15}{24}$

$\frac{4}{5} \times \frac{6}{7} = \frac{24}{35}$

$\frac{2}{3} \times \frac{6}{7} = \frac{4}{7}$

$\frac{7}{8} \div \frac{2}{3} = \frac{21}{16}$

$\frac{1}{2} \times \frac{7}{8} = \frac{7}{16}$

$\frac{4}{5} \div \frac{6}{7} = \frac{14}{15}$

$\frac{3}{4} \div \frac{6}{7} = \frac{7}{8}$

There are other answers besides the ones listed above.

Answer Key

Challenge Extension 1

When or if a team finishes early, extra challenge sheets are available. They may be assigned to teams or individuals. If a team is doing this for homework or classwork, then any member who gets an answer wins team points.

1. Use the fractions and any symbol to create a problem whose answer matches an answer in the box. Earn points for each math problem that has a correct answer.
2. **You may not use a fraction more than one time in any one problem you create, but you may use that fraction again in other solutions**

Fractions: $\frac{1}{2}, \frac{1}{3}, \frac{1}{4}, \frac{1}{5}, \frac{1}{6}$

You must show your work on this page.

Circle the correct answer and draw a straight line from your answer to the answer in the box to gain points.

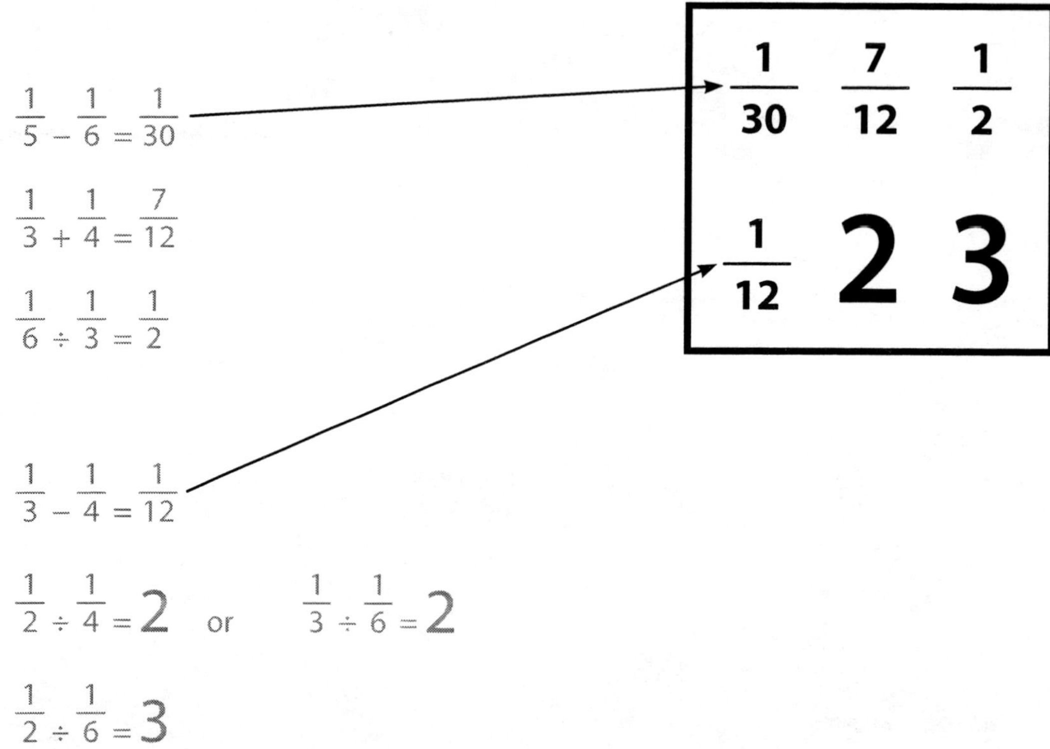

There are other answers besides the ones listed above.

Answer Key
Challenge Extension 2

Use the values given and any symbol to make a math problem that generates an answer on the math game board. Continue to make new math problems and circle the answers that you find. You win by circling answers that connect (vertically, horizontally, or diagonally) and get you from the start side to the finish side. You may not use the same value to create the denominator for both fractions.

All proper fraction answers are written in lowest terms. Circle your work that generates a correct answer and then circle that answer on the game board.

Values: 1, 2, 3, 4, 5, 6, 7, 8

$\frac{1}{5} \times \frac{7}{6} = \frac{7}{30}$ $\frac{2}{5} - \frac{1}{6} = \frac{7}{30}$

$\frac{2}{3} + \frac{4}{5} = \frac{22}{15}$

$\frac{3}{4} - \frac{2}{5} = \frac{7}{20}$

$\frac{4}{5} + \frac{1}{6} = \frac{29}{30}$

$\frac{1}{3} - \frac{2}{6} = \frac{2}{3}$

$\frac{4}{7} \times \frac{1}{2} = \frac{2}{7}$

$\frac{2}{4} + \frac{6}{8} = \frac{10}{8}$

$\frac{1}{2} + \frac{1}{3} = \frac{5}{6}$ $\frac{2}{3} \div \frac{4}{5} = \frac{5}{6}$

$\frac{1}{4} + \frac{2}{8} = \frac{1}{2}$

$\frac{7}{8} + \frac{1}{4} = \frac{9}{8}$

$\frac{3}{4} \times \frac{2}{5} = \frac{3}{10}$

$\frac{5}{7} - \frac{1}{6} = \frac{23}{42}$

$\frac{2}{5} \times \frac{1}{7} = \frac{2}{35}$

$\frac{6}{8} \div \frac{1}{2} = \frac{12}{8}$

$\frac{1}{4} + \frac{5}{6} = \frac{13}{12}$

$\frac{3}{4} + \frac{2}{8} = 1$

Start Side

$\frac{7}{30}$	$\frac{22}{15}$	$\frac{7}{20}$	$\frac{29}{30}$
$\frac{2}{3}$	$\frac{2}{7}$	$1\frac{1}{4}$	$\frac{5}{6}$
$\frac{1}{2}$	$\frac{9}{8}$	$\frac{3}{10}$	$\frac{23}{42}$
$\frac{2}{35}$	$\frac{12}{8}$	$\frac{13}{12}$	1

Finish Side

Answer Key
Challenge Extension 3

Choose two out of the three fractions and use one symbol to get the target answer. Solve the problem by writing your new problem below the three fraction choices.

$\frac{9}{2}$ $\frac{3}{4}$ $\frac{5}{9}$ $= \frac{7}{36}$

$\frac{3}{4} - \frac{5}{9} = \frac{7}{36}$

$\frac{11}{12}$ $\frac{1}{4}$ $\frac{1}{8}$ $= 2$

$\frac{1}{4} \div \frac{1}{8} = 2$

$\frac{3}{5}$ $\frac{8}{10}$ $\frac{6}{8}$ $= \frac{14}{10}$

$\frac{3}{5} + \frac{8}{10} = \frac{14}{10}$

$\frac{2}{7}$ $\frac{3}{5}$ $\frac{1}{4}$ $= \frac{31}{35}$

$\frac{2}{7} + \frac{3}{5} = \frac{31}{35}$

Answer Key
Individual Assessment: Addition

Problems are listed from easy to hard.

Use each whole number and each fraction once. Place them in the boxes to create a correct solution to each problem.

Values: 2, 3, 4, 6 $\frac{1}{2}$ $\frac{4}{7}$ $\frac{3}{9}$ $\frac{4}{5}$

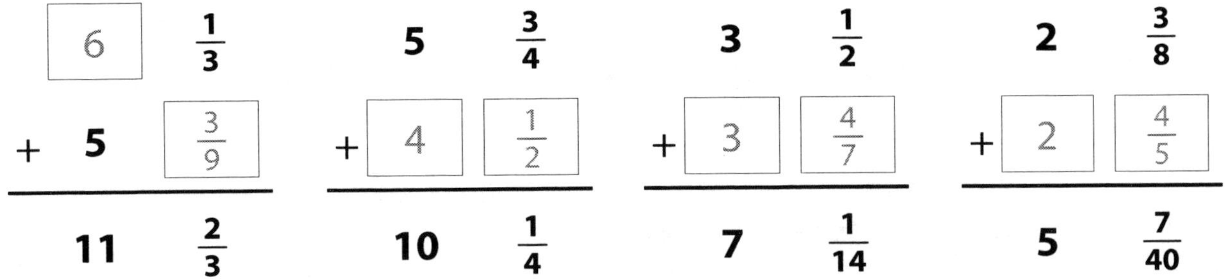

$$6\tfrac{1}{3} + 5\tfrac{3}{9} = 11\tfrac{2}{3}$$

$$5\tfrac{3}{4} + 4\tfrac{1}{2} = 10\tfrac{1}{4}$$

$$3\tfrac{1}{2} + 3\tfrac{4}{7} = 7\tfrac{1}{14}$$

$$2\tfrac{3}{8} + 2\tfrac{4}{5} = 5\tfrac{7}{40}$$

Answer Key
Individual Assessment: Subtraction

Problems are listed from easy to hard.

Use each whole number and each fraction once. Place them in the boxes to create a correct solution to each problem.

Values: 1, 1, 2, 2, $\frac{1}{5}$ $\frac{1}{4}$ $\frac{1}{3}$ $\frac{2}{3}$

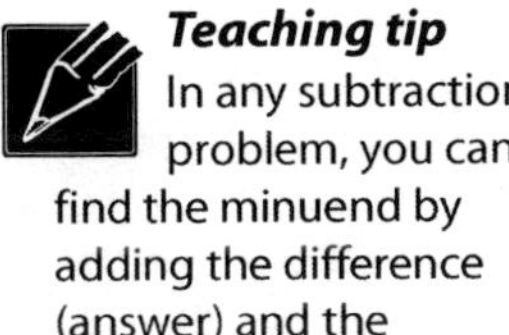

Teaching tip
In any subtraction problem, you can find the minuend by adding the difference (answer) and the subtrahend.

$$5\tfrac{3}{5} - 3\tfrac{1}{5} = 2\tfrac{2}{5}$$

$$5\tfrac{3}{8} - 4\tfrac{1}{3} = 2\tfrac{1}{24}$$

$$4\tfrac{1}{4} - 2\tfrac{1}{2} = 1\tfrac{3}{4}$$

$$3\tfrac{2}{6} - 1\tfrac{2}{3} = 1\tfrac{2}{3}\left(\tfrac{4}{6}\right)$$

Answer Key

Individual Assessment: Multiplication

Problems are listed from easy to hard.

Use four of the five values. Place one whole number in the each box to create a correct solution to each problem.

Values: 15, 16, 18, 20, 22

$\boxed{15} \times \frac{11}{11} = 15$

$\boxed{20} \times \frac{3}{4} = 15$

$\boxed{18} \times \frac{4}{6} = 12$

$\boxed{16} \times \frac{7}{8} = 14$

Use four of the five values. Place one value in a box to create a correct solution to each problem.

Values: 3, 4, 5, 6, 7

$$\frac{3}{6} \times \frac{6}{12} = \frac{1}{\boxed{4}}$$

$$10 \times \frac{\boxed{3}}{4} = 7\frac{1}{2}$$

$$\frac{\boxed{5}}{12} \times \frac{6}{7} = \frac{5}{14}$$

$$\frac{8}{9} \times \frac{\boxed{6}}{9} = \frac{16}{27}$$

Answer Key
Individual Assessment: Division

Teaching tip
Each division problem is asking, "How many of these (divisor) in that (dividend)?" See Student Guide, p. 10.

Problems listed from easy to hard.

Use four out of the five values. Place them in the boxes to create a correct solution to each problem.

Values: 1, 2, 3, 4, 5

$$4 \div \frac{\boxed{3}}{6} = 8$$

$$7 \div \frac{\boxed{1}}{7} = 49$$

$$\frac{9}{12} \div \frac{3}{\boxed{4}} = 1$$

$$\frac{3}{10} \div \frac{\boxed{2}}{5} = \frac{3}{4}$$

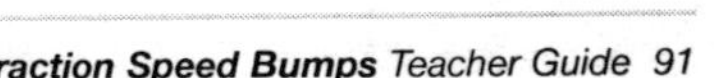

Answer Key

Individual Assessment: Thinking About Fractions

For use at any time during the unit.

Use the values provided to complete the following tasks:

Values: 1, 2, 3, 4, 5, 6, 7, 8, 9, 10, 11, 12

1. Write the three largest proper fractions:

   $\frac{11}{12}$ $\frac{10}{11}$ $\frac{9}{10}$

2. Write the three smallest fractions:

   $\frac{1}{12}$ $\frac{1}{11}$ $\frac{1}{10}$

3. Write three fractions equivalent to $\frac{1}{2}$:

   $\frac{2}{4}$ $\frac{3}{6}$ $\frac{4}{8}$ $\frac{5}{10}$ $\frac{6}{12}$

4. Write two fractions equivalent to $\frac{3}{4}$:

   $\frac{6}{8}$ $\frac{9}{12}$

5. Write two fractions that are closest to $\frac{1}{2}$ (but not equivalent to $\frac{1}{2}$):

   $\frac{5}{11}$ $\frac{6}{11}$

6. Write the three fractions that can be reduced to $\frac{1}{3}$:

   $\frac{2}{6}$ $\frac{3}{9}$ $\frac{4}{12}$

Answer Key

Individual Assessment: Comparing Fractions

For use at any time during the unit.

You should be able to study these and circle the largest fraction without doing all of the math work of changing to a common denominator.

1. $\frac{6}{10}$ (circled) or $\frac{1}{2}$ — $\frac{5}{10}$ is half… $\frac{6}{10}$ is more
2. $\frac{1}{2}$ (circled) or $\frac{21}{45}$ — $\frac{22.5}{45}$ is half
3. $\frac{21}{45}$ (circled) or $\frac{15}{63}$ — $\frac{15}{63}$ is close to $\frac{1}{4}$; $\frac{21}{45}$ is close to $\frac{1}{2}$
4. $\frac{15}{63}$ (circled) or $\frac{10}{80}$ — $\frac{15}{63}$ is close to $\frac{1}{4}$; $\frac{10}{80}$ is $\frac{1}{8}$
5. $\frac{15}{60}$ (circled) or $\frac{15}{63}$ — $\frac{15}{60}$ because 60ths are bigger pieces
6. $\frac{125}{500}$ (circled) or $\frac{12}{50}$ — $\frac{12}{50} = \frac{120}{500}$
7. $\frac{2000}{4001}$ or $\frac{2}{4}$ (circled) — $\frac{2}{4} = \frac{2000}{4000}$
8. $\frac{16}{33}$ or $\frac{5}{8}$ (circled) — $\frac{16}{33}$ is about $\frac{1}{2}$ … $\frac{5}{8}$ is more
9. $\frac{22}{66}$ (circled) or $\frac{11}{55}$ — $\frac{22}{66} = \frac{1}{3}$; $\frac{11}{55} = \frac{1}{5}$
10. $\frac{35}{101}$ (circled) or $\frac{11}{44}$ — $\frac{35}{101}$ is about $\frac{1}{3}$; $\frac{11}{44} = \frac{1}{4}$

Choose four problems you know you have circled correctly and write a simple explanation detailing how you solved each one.

Answer Key
Individual Assessment: Challenge

Use at the end of the unit.

1. Problem one:

   Values: 3, 5, 7, 9

   Choose to either multiply or divide. You may use each value only once in any problem to create proper fractions.

   Create the largest answer and the smallest answer you can.

Smallest $\frac{3}{7} \times \frac{5}{9} = \frac{5}{21}$ or $\frac{5}{9} \times \frac{3}{7} = \frac{5}{21} = 0.238$

Largest $\frac{5}{7} \div \frac{3}{9} = \frac{45}{21} = 2\frac{3}{21}$

2. Problem two:

   If the values change to 4, 6, 8, 10...can you create a problem that will generate a larger answer than the one from the first problem (still using each value only once)? Can you create a problem that will generate a smaller answer than the one from the first problem?

   Show your work.

Smallest $\frac{4}{8} \times \frac{6}{10} = \frac{3}{10}$ or $\frac{4}{10} \times \frac{6}{8} = \frac{3}{10} = 0.3$

Largest $\frac{6}{8} \div \frac{4}{10} = \frac{15}{8} = 1\frac{7}{8}$

Speed Bump Template

Small: 5 points; minimum spacing 4 cm

Medium: 7 points; minimum spacing 4.5 cm

Large: 3 points; minimum spacing 5.5 cm

Catching Tray Template

Master

Catching Tray

Fold on the dashed lines so there is a 1 cm edge (90 degrees) on three sides of the tray. Tape the corners.

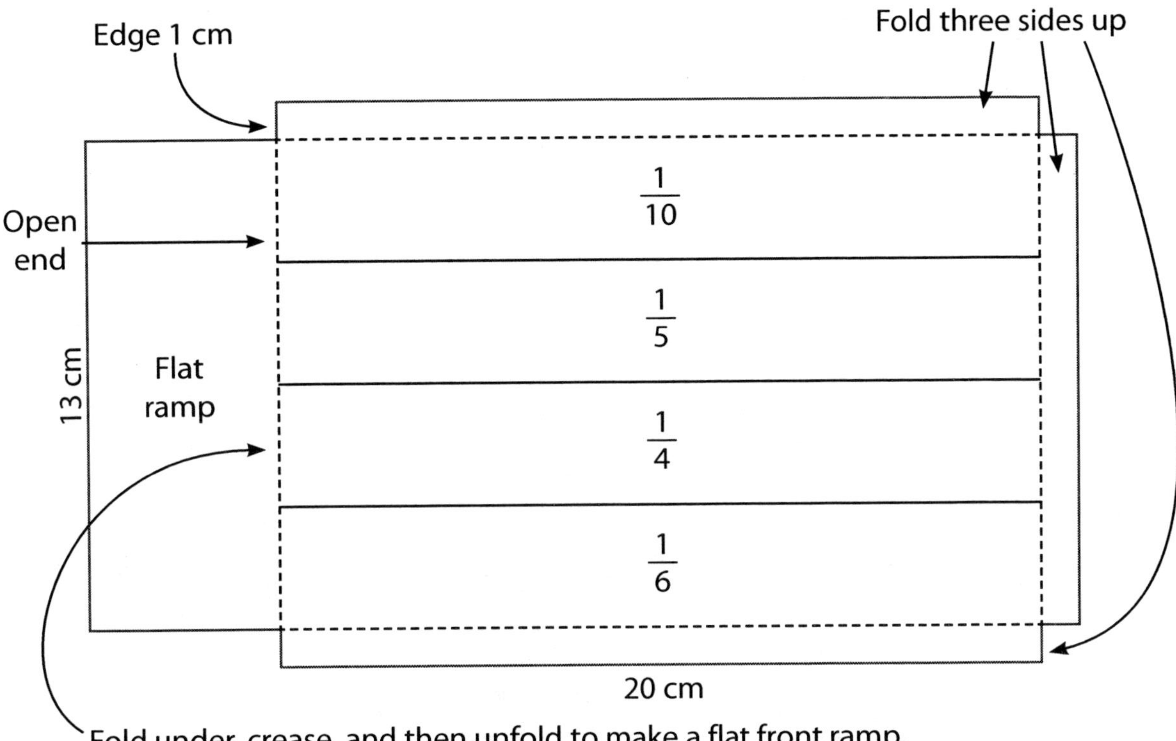

Fold the flat ramp under the box, crease the edge, and then unfold and tape the leading edge of the ramp down flat. This will make a slight rise at that front fold on the catching tray and keep some of the marbles from rolling off the tray. Even though the back edge of the tray is taped at the corners, some marbles will roll out the back end of the tray. Many marbles will hit the back edge, and then rebound and roll off the tray.

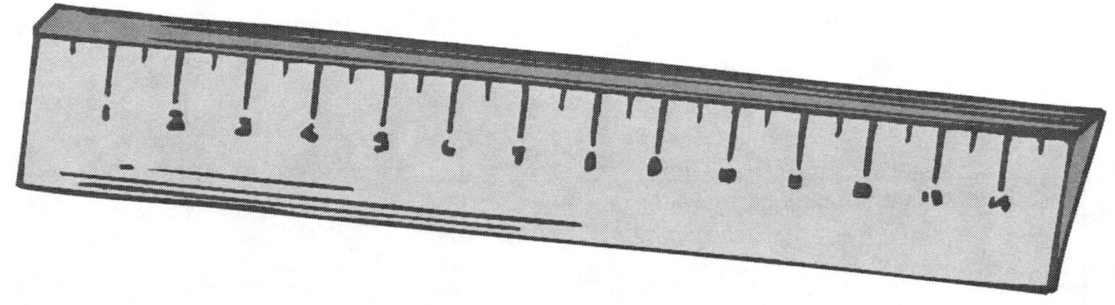

Speed Bump Trap Template

Fold sides and tape corners.

Cooperative Group Work Rubric

4

Exemplary

You consistently and actively help your group achieve its goals by communicating well with other group members, encouraging the group to work together, and willingly accepting and completing the necessary work of your daily role.

3

Expected

You usually help your group achieve its goals by communicating with other group members, encouraging your group to work together, and accepting and completing the necessary work of your role.

(If your evaluation is less than **EXPECTED,** try to use your cooperating skills more consistently.)

2

You sometimes help your group achieve its goals.

1

You do very little to help your group achieve its goals.

Fraction Score Sheet

Team name ______________________________

Date	Activity	Score	Bonus	Penalty	Points used	Total

Name: ______________________ Date: ______________

Pretest/Posttest

Skip the ones you do not know and work on the ones you think you know.

For numbers 1 through 5, circle the answers you choose.

1. Which two fractions are the largest?

   $\frac{9}{12}$ $\frac{7}{10}$ $\frac{2}{20}$ $\frac{3}{13}$ $\frac{5}{7}$ $\frac{4}{8}$ $\frac{4}{5}$ $\frac{4}{7}$

2. Which two fractions when added will give the smallest answer?

   $\frac{2}{13}$ $\frac{1}{11}$ $\frac{4}{40}$

3. Select two fractions that will give the largest answer when subtracted.

   $\frac{1}{7}$ $\frac{3}{8}$ $\frac{4}{5}$ $\frac{9}{10}$

4. Which two of these can be multiplied to get an answer less than $\frac{1}{2}$?

   $\frac{7}{8}$ $\frac{3}{5}$ $\frac{2}{3}$ $\frac{19}{20}$

5. How many of these can you use as divisors to get answers less than $\frac{1}{4}$?

   $\frac{1}{3}$ divided by… $\frac{1}{2}$ $\frac{1}{4}$ 2

6. Reduce these fractions and explain how you reduced them.

   a. $\frac{16}{32}$

   b. $\frac{40}{90}$

   c. $\frac{7}{21}$

7. Fill in the missing values.

   $$\frac{\square}{5} = \frac{6}{\square} = \frac{\square}{20} = \frac{200}{500}$$

8. Complete this adding problem:

$$\frac{2}{3} + \frac{\square}{8} = \frac{22}{\square}$$

9. Use these numbers: 4 5 12 16

   Select two and write the smallest fraction:

10. Use these numbers: 12 15 19 21

    Select two and write the largest fraction:

11. Complete this subtraction problem:

$$\begin{array}{r} 7\frac{\square}{70} \\ -\ 3\frac{8}{16} \\ \hline 4 \end{array}$$

Name: ______________________ Date: ____________

Teamwork Practice: Day 2

Place the values in the boxes to create the highest final answer. The answer to each problem goes in the circle and slides to the next circle to begin another math problem. All four symbols (+ – × ÷) go in the diamonds. Your team's best final answer will be added to your team's score sheet.

1. The class must use these five values: 6, 7, 8, 9, 10

2. This time, no symbol may be placed in the same diamond position as in the first sequence. Work for the highest answer. Your team's best final answer will be added to your team's score sheet.

3. This time the opposing team's Setter will place two different symbols in two of your diamonds before your team may begin. Your team places the other two symbols and all four values in the boxes. Try to get the highest answer. Your team's best final answer will be added to your team's score sheet.

Name: ______________________ Date: __________

Fractions Speed Bump: Day 3

Use the values provided and fill in the boxes to make the problem correct. Each value may be used only once. Earn the most points for being the first team to finish correctly. Teams all earn points but gain fewer points than those who have already finished correctly.

Values: 1, 2, 4, 5, 6, 8

$$\frac{4}{\square} = \frac{\square}{2} = \frac{\square}{\square} = \frac{3}{\square} = \frac{\square}{10}$$

Values: ____, ____, ____, ____, ____, ____

$$\frac{\square}{\square} = \frac{\square}{24} = \frac{\square}{30} = \frac{\square}{36} = \frac{\square}{42} = \frac{\square}{12}$$

Values: ____, ____, ____, ____, ____, ____

$$\frac{\square}{\square} + \frac{\square}{8} + \frac{\square}{10} = \frac{22}{40} = \frac{\square}{\square}$$

Name: ______________________________ Date: ______________

Adding Fractions Speed Bump: Day 4

Use each value once. Place the values in the boxes so that you create a correct mathematical sentence.

1. **Values: 1, 2, 3, 4, 5, 8**

$$\frac{\square}{3} + \frac{\square}{\square} + \frac{\square}{6} = \frac{20}{12} = 1\frac{\square}{12} = 1\frac{2}{\square}$$

2. **Values:** ____, ____, ____, ____, ____, ____

$$\frac{\square}{\square} + \frac{\square}{\square} + \frac{3}{\square} = \frac{72}{60} = 1\frac{\square}{60} = 1\frac{1}{5}$$

3. Use these values to find the largest possible answer. Use proper fractions only. You may not use whole numbers ($\frac{3}{1}$).

Values: ____, ____, ____, ____, ____, ____, ____

$$\frac{3}{\square} + \frac{\square}{\square} + \frac{\square}{\square} = \frac{\square}{\square} = 2\frac{1}{12}$$

4. Use the values to find the largest answer and the smallest answer. Use proper fractions only. You may not use any whole number designations ($\frac{3}{1}$). Discuss a team strategy before beginning. Points are awarded to teams that finish quickly. Additional points are awarded if your team has the correct largest and/or smallest answer.

Values: ____, ____, ____, ____, ____, ____

$$\frac{\square}{\square} + \frac{\square}{\square} + \frac{\square}{\square} = \frac{\square}{\square}$$ largest

Name: ______________________________ Date: ____________

Subtracting Fractions Speed Bump: Day 6

1. Use only four of the eight values given. Create two proper fractions that when subtracted will give the **largest** answer available. If finished early, go for the bonus assignment. ***Bonus:*** Use the other four values (not yet used in the first challenge) and find the smallest answer you can. Team(s) with the **smallest** answer will gain bonus points to add to their total.

Values: 2, 4, 5, 6, 8, 10, 15, 20 ***Bonus:***

$\frac{\square}{\square} - \frac{\square}{\square} = \square$ $\frac{\square}{\square} - \frac{\square}{\square} = \square$

2. Use all of the values once. Use four of the values to create two proper fractions to use in the first subtraction problem. After solving the problem, use that answer and subtract another proper fraction by using the remaining two values. Use experimentation to find the **smallest** answer available. No negative answers allowed.

Values: ____, ____, ____, ____, ____, ____

$\frac{\square}{\square} - \frac{\square}{\square} = \bigcirc$

$\bigcirc - \frac{\square}{\square} = \square$

3. Subtract, then add. This is a challenge to build the **largest** answer possible. Use all of the values once. Use four of the values to create two proper fractions to use in the subtraction problem. Use the two remaining values to make another proper fraction, that when added to the first answer will give the **largest** total.

Values: ____, ____, ____, ____, ____, ____

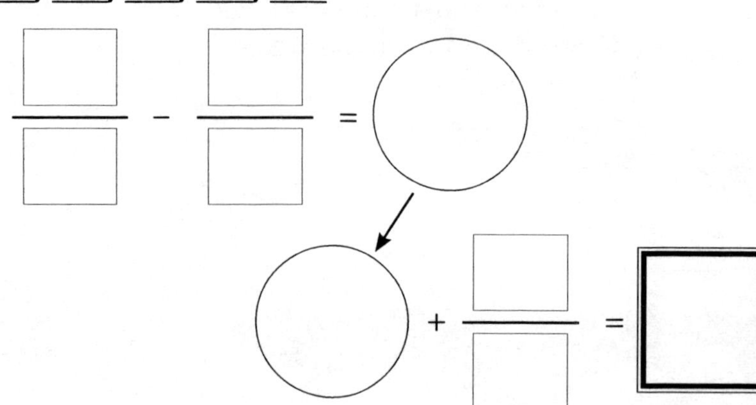

Name: ______________________ Date: ____________

Multiplying Fractions Speed Bump: Day 7

1. **First challenge:** Use the values and experiment with multiplying **proper fractions**. Search for the largest and the smallest answer. After working on the challenge, your team will need to decide if it wants to commit to having the largest or smallest answer. You must circle the word "largest" or "smallest" on your paper. You gain points if your selected answer is the largest or the smallest. If only a few teams match the correct answer (large or small) the points earned will be greater. Many correct answers to "largest" or "smallest" will reduce the points earned. Good luck choosing and working correctly.

Values: 2, 4, 6, 8, 10, 12

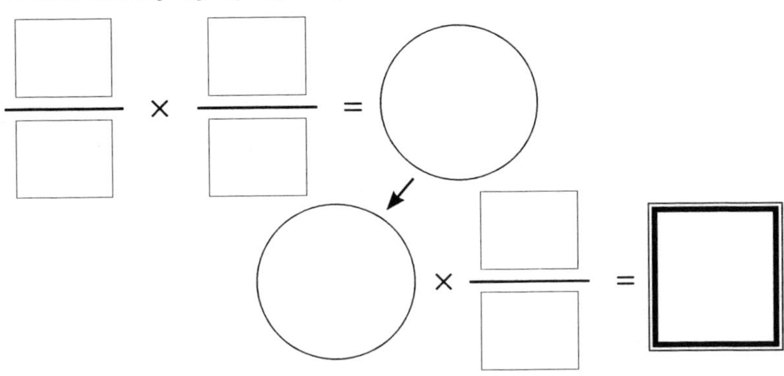

Largest Smallest

2. **Second challenge:** Use the same values as provided above. First your opponent's Setter places *values* in any two of your team's fraction squares and they cannot be moved. You will only use two of the four values that remain. Try to generate only a **large** answer. The larger answers score more points. **Proper fractions only.**

3. **Third challenge:** Continue using these six values (**2, 4, 6, 8, 10, 12**). This time, your opponent's Setter places four of the given *values* in any of the six squares in the problem below. Your team may choose to replace one of the "placed" values with either of the two remaining values. You do not have to change a value. Again, try to generate a **large** answer. If your answer is greater than your opponent's answer, you win. You also win if you have the largest answer in the class. **Improper fractions allowed** for this challenge.

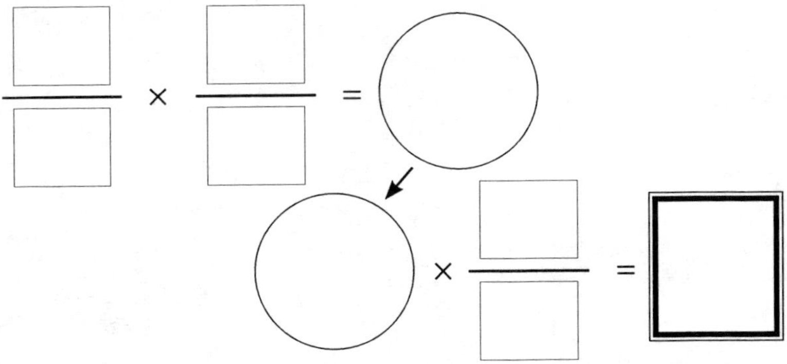

Name: ______________________ Date: ____________

Dividing Fractions Speed Bump: Day 9

1. Use the four values to create the division problem that gives the **largest** answer for the two challenges below. When you are finished try the bonus challenge. For the "bonus," points will be awarded for the answer closest to $1\frac{7}{16}$.

Values: 2, 4, 6, 8—To be used with the three challenges below.

Proper fractions only:

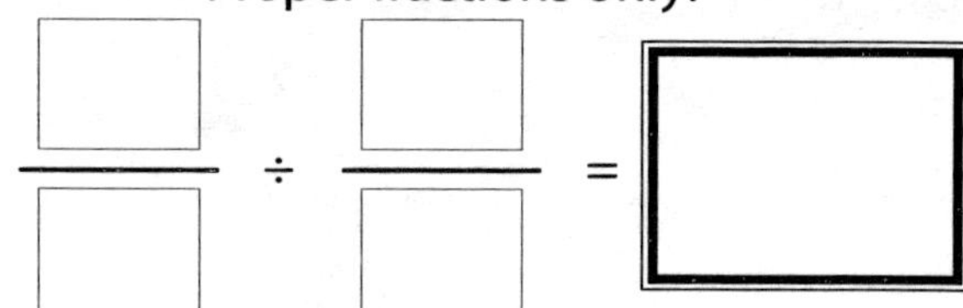

Improper fractions allowed:

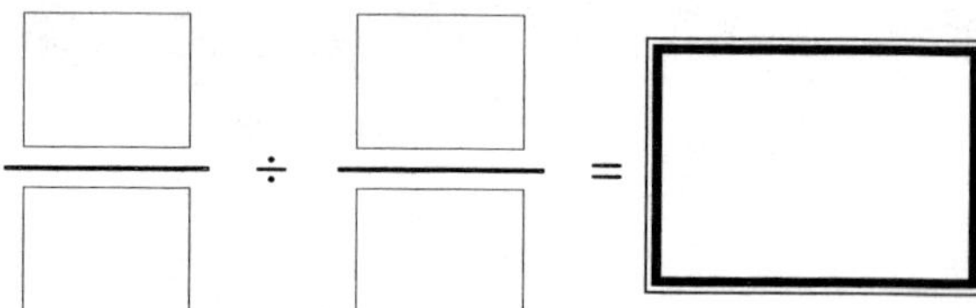

Bonus:

2. Use the six values to create the **smallest** answer. If you believe you have the smallest answer you may try for the bonus. If you do have the smallest answer, then you win bonus points, but if another team has a smaller answer, then you lose some points. Circle the word "smallest" below if you want to compete for the bonus points.

Values: ____, ____, ____, ____, ____, ____

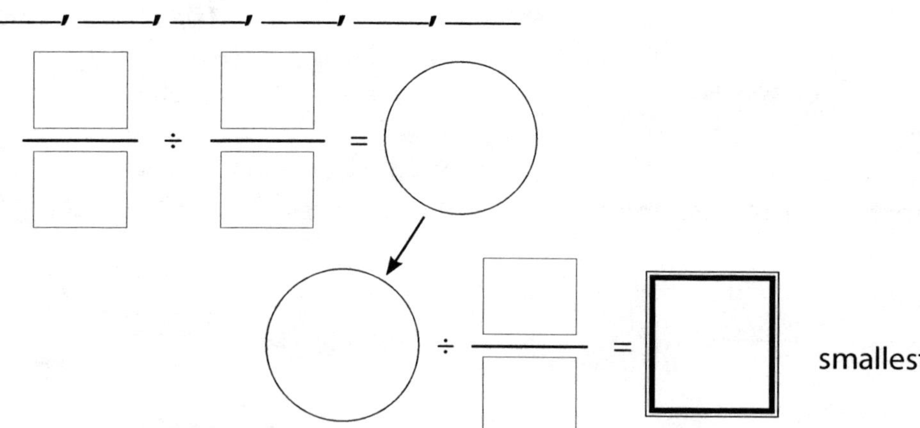

3. Try for the **largest or the smallest** answer this time. Use the six values given to create your answer. Your opponent's Setter will decide the placement of the two different symbols (+ - x ÷, placed in the diamonds) and you may not move them. All teams that find either the largest or the smallest answer will gain points. All ties earn points.

Values: 3, 5, 7, 8, 9, 10

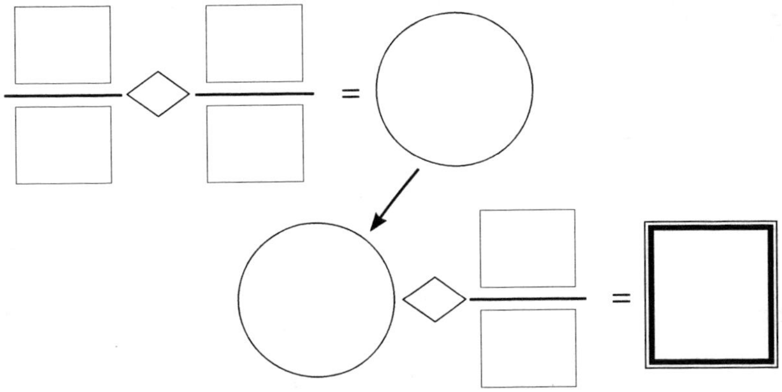

4. Second sequence: Your opponent's Setter must pick a different combination of two different symbols (+ – × ÷). Use the same values again. All teams that find either the largest or the smallest answer will gain points. All ties earn points.

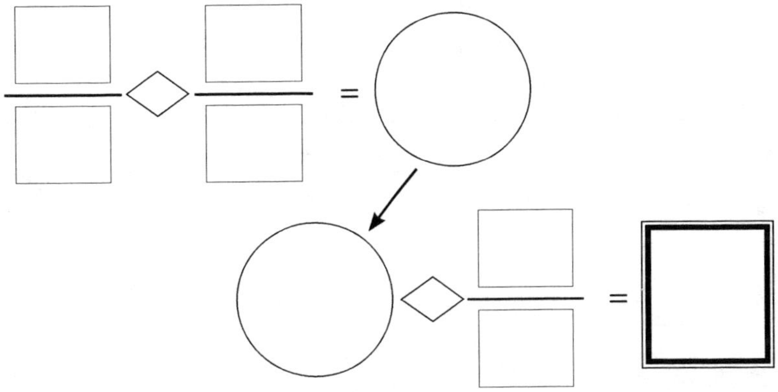

Name: ______________________ Date: ____________

Addition Extension Activity 1e

Place the whole numbers and fractions provided in the boxes to create the answer indicated.

Whole numbers: 2, 5, 7, 8 Fractions: $\frac{1}{3}$, $\frac{2}{8}$, $\frac{3}{4}$, $\frac{4}{9}$

You may use any fraction or any whole number twice in any problem.

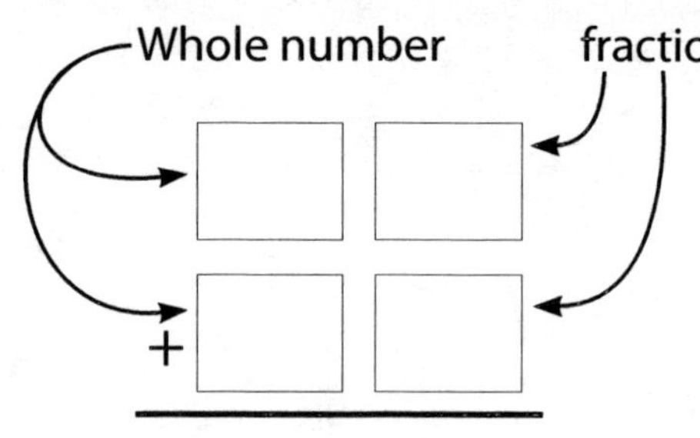

$13\ \frac{2}{3}$

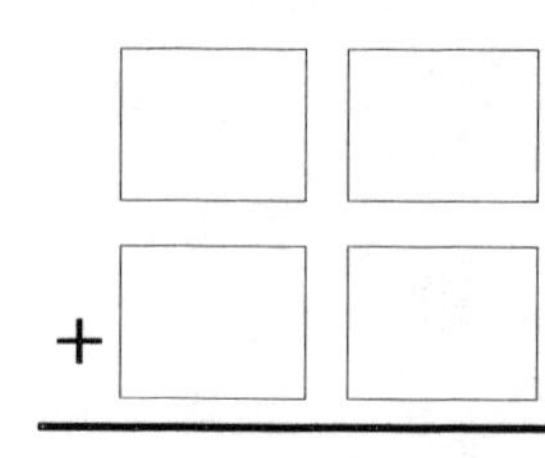

$4\ \frac{8}{9}$

+

$17\ \frac{1}{2}$

+

$7\ \frac{7}{9}$

+

$12\ \frac{50}{72}$

+

16

+

$10\ \frac{13}{12}$

+

$9\ \frac{14}{24}$

Name: ______________________________ Date: ______________

Addition Extension Activity 1h

Place the whole numbers and fractions provided in the boxes to create the answer indicated.

Whole numbers: 2, 5, 7, 8   Fractions: $\frac{1}{3}$, $\frac{2}{8}$, $\frac{3}{4}$, $\frac{4}{9}$

You <u>may</u> use any fraction or any whole number twice in any problem.

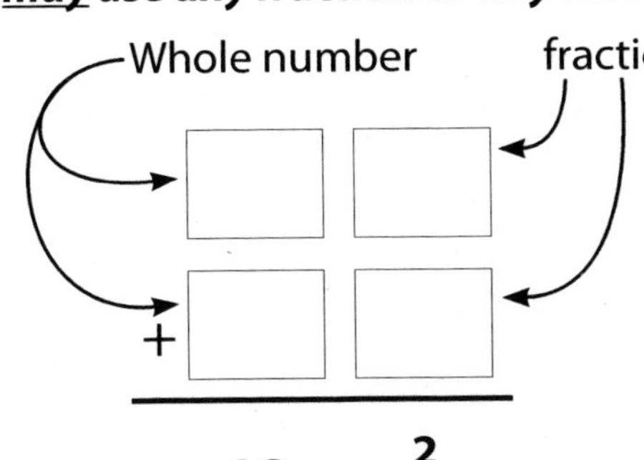

13 $\frac{2}{3}$

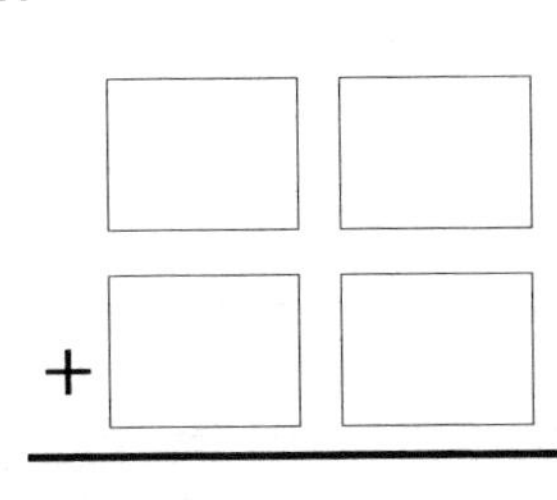

4 $\frac{8}{9}$

+

17 $\frac{1}{2}$

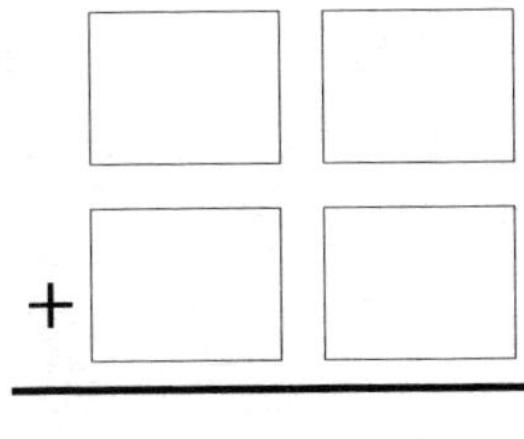

7 $\frac{7}{9}$

+

12 $\frac{25}{36}$

+

16

+

11 $\frac{1}{12}$

+

9 $\frac{7}{12}$

Name: ______________________________ Date: ______________

Subtraction Extension Activity 1e

Place the whole numbers and fractions provided in the boxes to create the answer indicated.

Whole numbers: 3, 4, 6, 9 Fractions: $\frac{2}{10}$, $\frac{4}{6}$, $\frac{9}{12}$, $\frac{5}{10}$

You may <u>not</u> use any fraction or any whole number twice in any one problem.

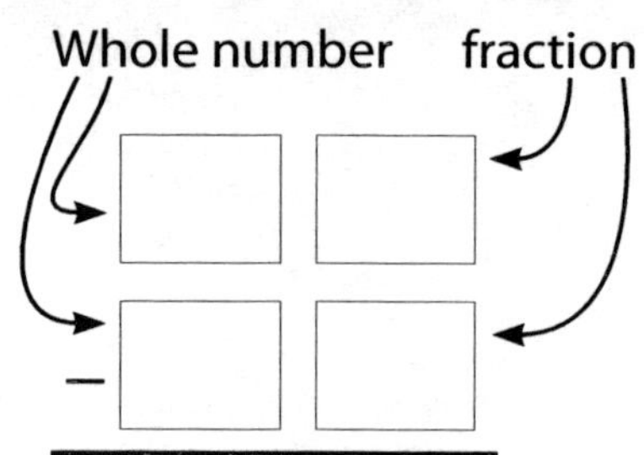

$1\ \frac{3}{10}$

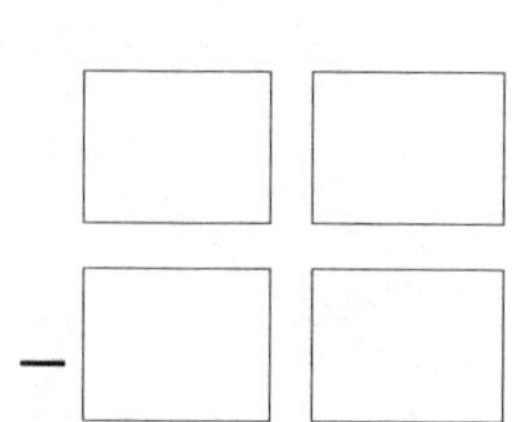

$2\ \frac{1}{12}$

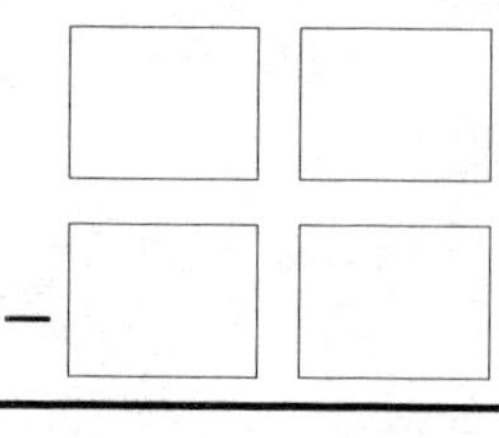

$3\ \frac{14}{30}$

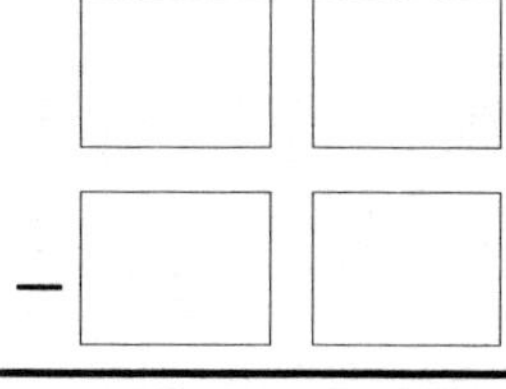

$3\ \frac{15}{60}$,

$5\ \frac{33}{60}$

$5\ \frac{7}{10}$

$2\ \frac{11}{12}$

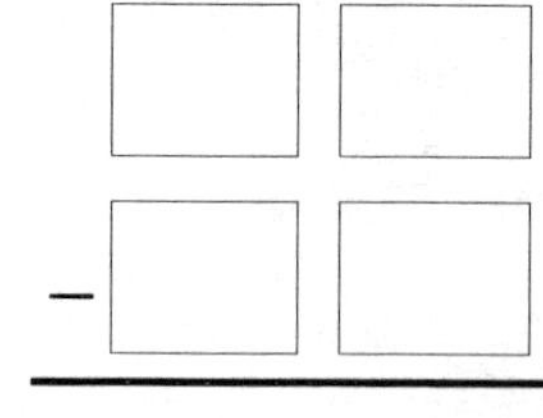

$4\ \frac{16}{30}$

Name: ____________________ Date: ____________

Subtraction Extension Activity 1h

Place the whole numbers and fractions provided in the boxes to create the answer indicated.

Whole numbers: 3, 4, 6, 9 **Fractions: $\frac{2}{10}$, $\frac{4}{6}$, $\frac{9}{12}$, $\frac{5}{10}$**

You may <u>not</u> use any fraction or any whole number twice in any one problem.

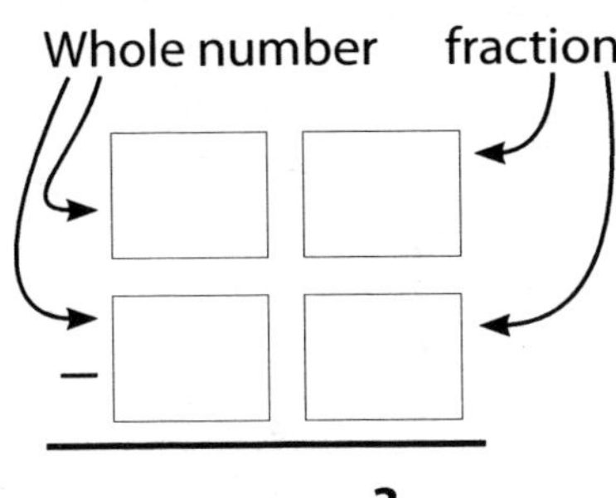

$1\ \frac{3}{10}$

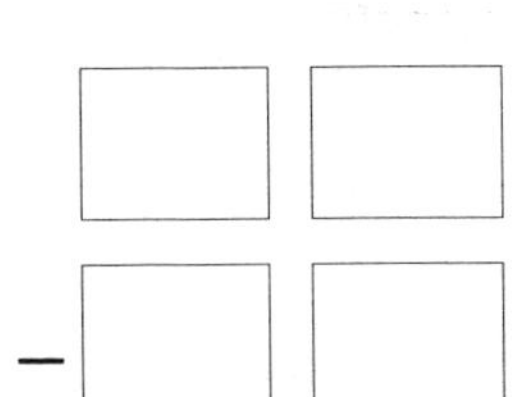

$2\ \frac{1}{12}$

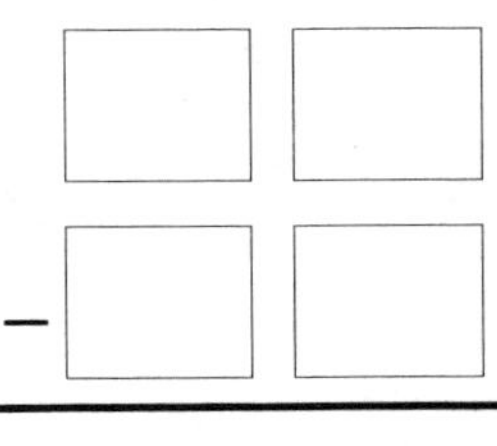

$3\ \frac{7}{15}$

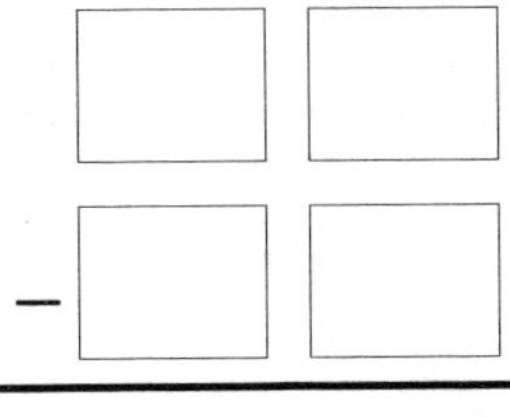

$3\ \frac{1}{4}$

$5\ \frac{11}{20}$

$5\ \frac{7}{10}$

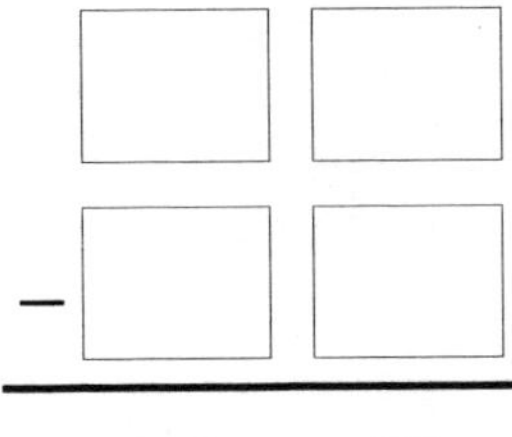

$2\ \frac{11}{12}$

$4\ \frac{8}{15}$

Name: ______________________________ Date: ______________

Addition and Subtraction Extension Activity

Add and subtract only.

Use the fractions provided to generate answers that are on the game board. Any three in a row wins points, but the more answers you find, the higher your point total. On this paper, write a math problem that gives an answer matching a fraction on the game board. Put an "x" over the matching fraction on the game board. Then write a number "1" by your first answer and by the "x"ed-out answer on the game board. Continue numbering your fraction problems and answers that match using this method. Fractions can be used repeatedly in any combination.

Fractions: $\frac{1}{2}, \frac{1}{3}, \frac{1}{4}, \frac{1}{5}, \frac{1}{6}, \frac{1}{7}, \frac{1}{8}$

$\frac{1}{6}$	$\frac{7}{10}$	$\frac{11}{24}$
$\frac{12}{35}$	$\frac{5}{6}$	$\frac{1}{12}$
$\frac{7}{12}$	$\frac{5}{12}$	$\frac{3}{8}$

Name: ______________________________ Date: ______________

Multiplication Extension Activity

Use the values given and fill in the boxes so that you create a correct math problem.

Each number listed is used once. Proper fractions only.

3, 3, 4, 5, 9, 20

$$\frac{\square}{\square} \times \frac{\square}{\square} = \frac{\square}{\square}$$

1, 2, 3, 4, 4, 6, 12, 24

$$\frac{\square}{\square} \times \frac{\square}{\square} = \frac{\square}{\square} = \frac{\square}{\square}$$

2, 3, 5, 6, 8, 10, 40, 60

$$\frac{\square}{\square} \times \frac{\square}{\square} = \frac{\square}{\square} = \frac{\square}{\square}$$

5, 7, 7, 30, 49, ____

$$\frac{\square}{\square} \times \frac{\square}{\square} = \frac{\square}{\square}$$

Name: ______________________________ Date: ______________

Multiplying With Fractions Extension

If your multiplier is a proper fraction, your answer will be smaller than the number you started with. When you multiply using a fraction you get *part* of your original value. Your answer is less than your original value.

Start with 16 and multiply by $\frac{1}{2}$.

You are really asking, "What is half of sixteen?"

Original value	× the fraction $\frac{1}{2}$	= answer
16	$\frac{1}{2}$	
8	$\frac{1}{2}$	
4	$\frac{1}{2}$	
2	$\frac{1}{2}$	
1	$\frac{1}{2}$	
$\frac{1}{2}$	$\frac{1}{2}$	
$\frac{1}{4}$	$\frac{1}{2}$	
$\frac{1}{8}$	$\frac{1}{2}$	
$\frac{1}{16}$	$\frac{1}{2}$	

Notice that multiplying by $\frac{1}{2}$ gives the same answer as dividing by 2. See Student Guide, page 10.

Use the information from the table on the previous page and complete the tables below:

16		
8		
4		$\frac{1}{4}$
2		
1		
$\frac{1}{2}$		
$\frac{1}{4}$		
$\frac{1}{8}$		
$\frac{1}{16}$		

$\frac{2}{3}$		
		$\frac{1}{144}$

1. What patterns did you notice in the tables above?

2. How can what you have observed in the activities above make you work faster and more accurately?

Name: ______________________________ Date: ______________

Division Extension

1, 2, 4, 12, 24

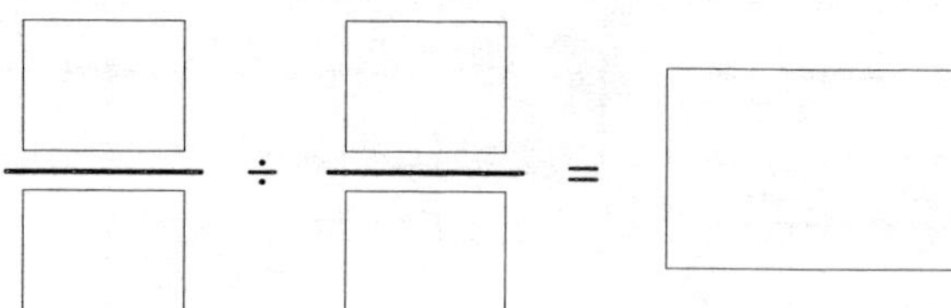

1, 3, 3, 8, 8

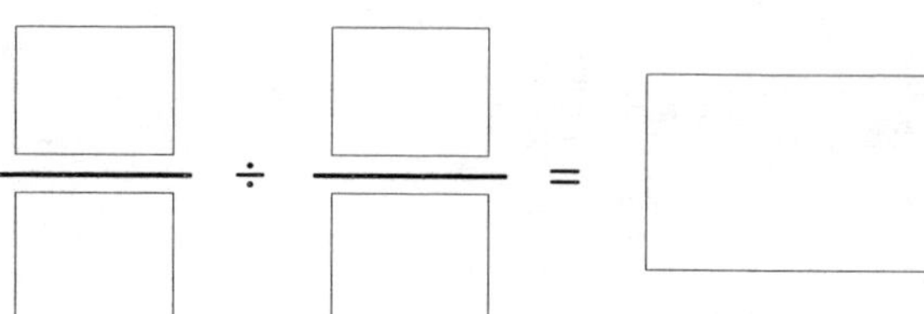

2, 3, 8, 10—These are the values for the division problem only. You must provide the other numbers as you figure out the problem.

$\frac{\square}{\square} \div \frac{\square}{\square} = \frac{\square}{\square} = \square\frac{\square}{\square} = 1\frac{1}{5}$

1, 1, 1, 4, 8, 32

$\frac{\square}{\square} \div \frac{\square}{\square} = \frac{\square}{\square}$

Name: ____________________ Date: __________

Dividing With Fractions Extension

When dividing by a fraction your answer will be larger if the original value (dividend) is a whole number. When dividing a fraction by a smaller fraction, your answer will be larger than the dividend. When dividing a fraction by a larger fraction, the answer will be larger than the dividend.

Start with 10 and divide by $\frac{1}{2}$.

You are really asking, "How many one halves are in ten?" The answer is twenty.

Start with $\frac{1}{2}$ and divide by $\frac{1}{4}$.

You are really asking, "How many one fourths are in one half?" The answer is two.

Start with $\frac{1}{4}$ and divide by $\frac{1}{2}$.

You are really asking, "How many one halves are in one fourth?" The answer is one half. It takes one half of a half to make one fourth.

Original value ÷ the fraction $\frac{1}{2}$ = answer

Original value	the fraction	answer
16	$\frac{1}{2}$	
8	$\frac{1}{2}$	
4	$\frac{1}{2}$	
2	$\frac{1}{2}$	
1	$\frac{1}{2}$	
$\frac{1}{2}$	$\frac{1}{2}$	
$\frac{1}{4}$	$\frac{1}{2}$	
$\frac{1}{8}$	$\frac{1}{2}$	
$\frac{1}{16}$	$\frac{1}{2}$	

When dividing, you are always asking, "How many of this fraction does it take to make this value?"

Use the information from the table on the previous page and complete the tables below.

16		
8		
4		
2		**16**
1		
$\frac{1}{2}$		
$\frac{1}{4}$		
$\frac{1}{8}$		
$\frac{1}{16}$		

24		
		$\frac{3}{4}$

1. What patterns did you notice in the tables above?

2. How can what you have observed in the activities above make you work faster and more accurately?

3. Did the divisor in this problem change the pattern in any way?

4. Which column (line of values) in table three proves this statement: "Every fraction is a division problem."

Name: ______________________________ Date: ______________

Fraction to Decimal Practice Extension

Try to match each fraction with its decimal equivalent and then check your work using a calculator. Draw lines from each fraction to the decimal equivalent.

$\frac{4}{9}$	0.666
$\frac{9}{16}$	0.9090
$\frac{3}{8}$	0.555
$\frac{4}{6}$	0.75
$\frac{5}{9}$	0.444
$\frac{2}{5}$	0.5625
$\frac{7}{8}$	0.375
$\frac{10}{11}$	0.1428
$\frac{3}{4}$	0.875
$\frac{1}{7}$	0.40

Name: ____________________ Date: __________

Fraction to Decimal Extension

Use the values given to fill in the numerator box in each fraction so that you create a fraction equivalent to one of the decimal answers in the right-hand column. Draw a line from the newly created fraction to the proper decimal. Use each value once.

Values: 3, 9, 2, 10, 7, 4

$\frac{\square}{16}$	0.875
$\frac{\square}{4}$	0.40
$\frac{\square}{5}$	0.75
$\frac{\square}{11}$	0.444
$\frac{\square}{8}$	0.5625
$\frac{\square}{9}$	0.9090

Name: ______________________________ Date: ______________

Multiplication and Division Extension

Multiply and divide only

Use the fractions provided to generate answers that are on the game board. Any three in a row wins points, but the more answers you find, the higher your point total. On this paper write a math problem that gives an answer that matches a fraction on the game board. Put an "x" over the matching fraction on the game board. Then write a number "1" by your first answer and by the "x"ed-out answer on the game board. Continue numbering your fraction problems and answers that match using this method. Fractions can be used repeatedly in any combination.

Fractions: $\frac{1}{2}, \frac{2}{3}, \frac{3}{4}, \frac{4}{5}, \frac{5}{6}, \frac{6}{7}, \frac{7}{8}$

$\frac{4}{3}$	$\frac{2}{6}$	$\frac{15}{24}$
$\frac{24}{35}$	$\frac{4}{7}$	$\frac{21}{16}$
$\frac{7}{16}$	$\frac{14}{15}$	$\frac{20}{21}$

Name: ____________________ Date: ____________

Challenge Extension 1

When or if a team finishes early, extra challenge sheets are available. They may be assigned to teams or individuals. If a team is doing this for homework or classwork, then any member who gets an answer wins team points.

1. Use the fractions and any symbol to create a problem that has an answer that matches an answer in the box. Earn points for each math problem that has a correct answer.
2. **You may not use a fraction more than one time in any one problem you create, but you may use that fraction again in other solutions.**

Fractions: $\frac{1}{2}, \frac{1}{3}, \frac{1}{4}, \frac{1}{5}, \frac{1}{6}$

You must show your work on this page.

Circle the correct answer and draw a straight line from your answer to the answer in the box to gain points.

$\frac{1}{30}$	$\frac{7}{12}$	$\frac{1}{2}$
$\frac{1}{12}$	2	3

Name: ______________________________ Date: ______________

Challenge Extension 2

Use the values given and use any symbol to make a math problem that generates an answer on the math game board. Continue to make new math problems and circle the answers that you find. You win by circling answers that connect (vertically, horizontally, or diagonally) and get you from the start side to the finish side. You may not use the same value to create the denominator for both fractions.

All proper fraction answers are written in lowest terms. Circle your work that generates a correct answer and then circle that answer on the game board.

Values: 1, 2, 3, 4, 5, 6, 7, 8

Start Side

$\frac{7}{30}$	$\frac{22}{15}$	$\frac{7}{20}$	$\frac{29}{30}$
$\frac{2}{3}$	$\frac{2}{7}$	$1\frac{1}{4}$	$\frac{5}{6}$
$\frac{1}{2}$	$\frac{9}{8}$	$\frac{3}{10}$	$\frac{23}{42}$
$\frac{2}{35}$	$\frac{12}{8}$	$\frac{13}{12}$	1

Finish Side

Name: ______________________ Date: ____________

Challenge Extension 3

Choose two out of the three fractions and use one symbol to get the target answer. Solve the problem by writing your new problem below the three fraction choices.

$\frac{9}{2}$ $\frac{3}{4}$ $\frac{5}{9}$ $= \frac{7}{36}$

$\frac{11}{12}$ $\frac{1}{4}$ $\frac{1}{8}$ $= 2$

$\frac{3}{5}$ $\frac{8}{10}$ $\frac{6}{8}$ $= \frac{14}{10}$

$\frac{2}{7}$ $\frac{3}{5}$ $\frac{1}{4}$ $= \frac{31}{35}$

Name: ______________________ Date: __________

Individual Assessment: Addition

Use each whole number and each fraction once. Place them in the boxes to create a correct solution to each problem.

Values: 2, 3, 4, 6 $\frac{1}{2}$ $\frac{4}{7}$ $\frac{3}{9}$ $\frac{4}{5}$

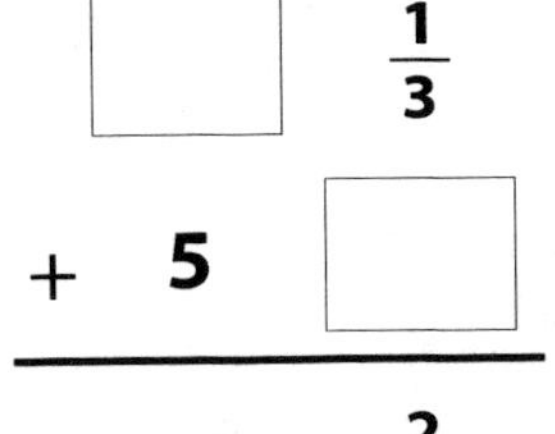

□ $\frac{1}{3}$	5 $\frac{3}{4}$	3 $\frac{1}{2}$	2 $\frac{3}{8}$
+ 5 □	+ □ □	+ □ □	+ □ □
11 $\frac{2}{3}$	10 $\frac{1}{4}$	7 $\frac{1}{14}$	5 $\frac{7}{40}$

Master

Name: ______________________ Date: __________

Individual Assessment: Subtraction

Use each whole number and each fraction once. Place them in the boxes to create a correct solution to each problem.

Values: 1, 1, 2, 2 $\frac{1}{5}$ $\frac{1}{4}$ $\frac{1}{3}$ $\frac{2}{3}$

5 $\frac{3}{5}$	5 $\frac{3}{8}$	4 □	3 $\frac{2}{6}$
− 3 □	− 4 □	− 2 $\frac{1}{2}$	− 1 □
□ $\frac{2}{5}$	□ $\frac{1}{24}$	□ $\frac{3}{4}$	□ $\frac{2}{3}$

Name: ______________________________ Date: ______________

Individual Assessment: Multiplication

Use four of the five values. Place one whole number in the each box to create a correct solution to each problem.

Values: 15, 16, 18, 20, 22

$\square \times \frac{11}{11} = 15$

$\square \times \frac{3}{4} = 15$

$\square \times \frac{4}{6} = 12$

$\square \times \frac{7}{8} = 14$

Use four of the five values. Place one value in a box to create a correct solution to each problem.

Values: 3, 4, 5, 6, 7

$\frac{3}{6} \times \frac{6}{12} = \frac{1}{\square}$

$10 \times \frac{\square}{4} = 7\frac{1}{2}$

$\frac{\square}{12} \times \frac{6}{7} = \frac{5}{14}$

$\frac{8}{9} \times \frac{\square}{9} = \frac{16}{27}$

Name: ______________________________ Date: ______________

Individual Assessment: Division

Use four out of the five values. Place them in the boxes to create a correct solution to each problem.

Values: 1, 2, 3, 4, 5

$$4 \div \frac{\square}{6} = 8$$

$$7 \div \frac{\square}{7} = 49$$

$$\frac{9}{12} \div \frac{3}{\square} = 1$$

$$\frac{3}{10} \div \frac{\square}{5} = \frac{3}{4}$$

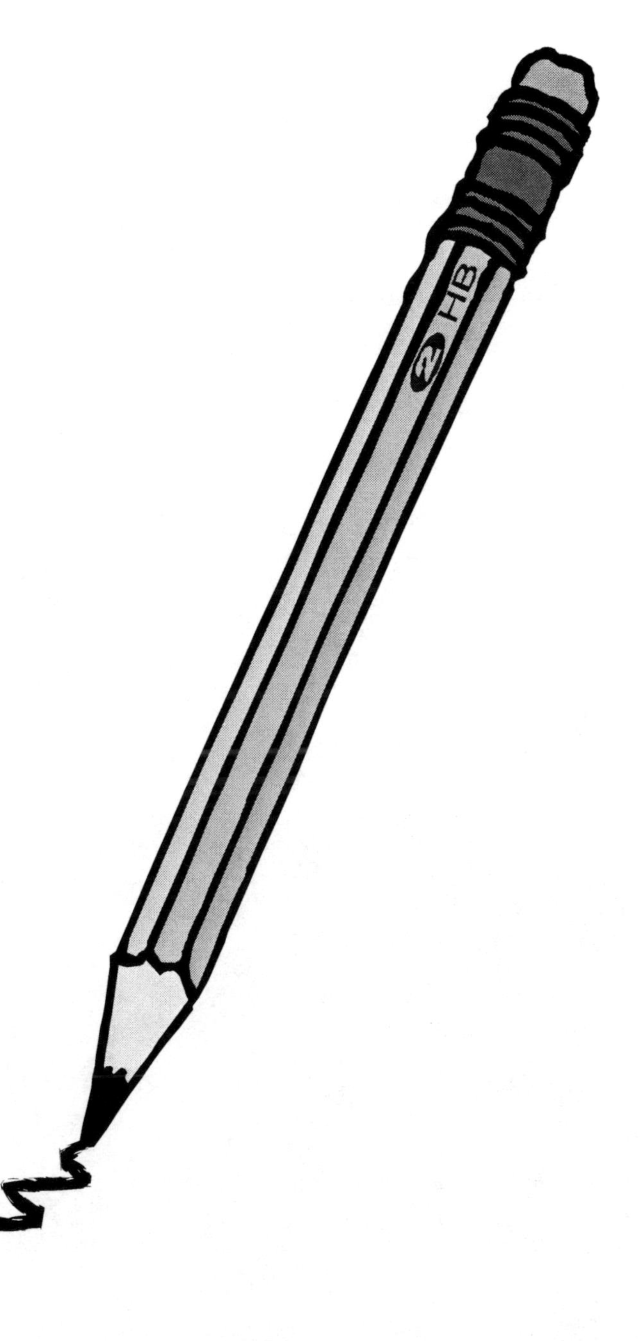

Name: ______________________________ Date: ______________

Individual Assessment: Thinking About Fractions

Use the values provided to complete the following tasks:

Values: 1, 2, 3, 4, 5, 6, 7, 8, 9, 10, 11, 12

1. Write the three largest proper fractions:

2. Write the three smallest fractions:

3. Write three fractions equivalent to $\frac{1}{2}$:

4. Write two fractions equivalent to $\frac{3}{4}$:

5. Write two fractions that are closest to $\frac{1}{2}$ (but not equivalent to $\frac{1}{2}$):

6. Write the three fractions that can be reduced to $\frac{1}{3}$:

Name: ______________________________ Date: ______________

Individual Assessment: Comparing Fractions

You should be able to study these and circle the largest fraction without doing all of the math work of changing to a common denominator.

1. $\frac{6}{10}$ or $\frac{1}{2}$

2. $\frac{1}{2}$ or $\frac{21}{45}$

3. $\frac{21}{45}$ or $\frac{15}{63}$

4. $\frac{15}{63}$ or $\frac{10}{80}$

5. $\frac{15}{60}$ or $\frac{15}{63}$

6. $\frac{125}{500}$ or $\frac{12}{50}$

7. $\frac{2000}{4001}$ or $\frac{2}{4}$

8. $\frac{16}{33}$ or $\frac{5}{8}$

9. $\frac{22}{66}$ or $\frac{11}{55}$

10. $\frac{35}{101}$ or $\frac{11}{44}$

Choose four problems you know you have circled correctly and write a simple explanation detailing how you solved each one.

Name: ______________________________ Date: ______________

Individual Assessment: Challenge

1. Problem one:

   Values: 3, 5, 7, 9

   Choose to either multiply or divide. You may use each value only once in any problem to create proper fractions.

   Create the largest answer and the smallest answer you can.

2. Problem two:

   If the values change to **4**, **6**, **8**, **10**… can you create a problem that will generate a larger answer than the one from the first problem (still use each value only once)? Can you create a problem that will generate a smaller answer than the one from the first problem?

   Show your work.

STUDENT GUIDE

Name: ____________________

FRACTION SPEED BUMPS

Introduction

Welcome to *Fraction Speed Bumps*. In the days ahead, you and your team will engage in friendly competition with other teams. You will compete in a variety of fraction challenge sheets and speed bump events. Speed bumps may slow you down, but if you work and plan as a team you will be successful. This guide provides you with some fraction-thinking tools. Use the tools to help your team conquer some of the challenges that you will encounter.

Your main challenge is to forge a team that can be successful by looking at fractions and then making them work for you. There will be two kinds of competition:

1. *Fraction speed bump challenges*

   Your team will try to complete the daily fraction math challenges. During the competition, you will select the best values and place them so that your team can succeed. Your job is to earn as many points as possible each day.

2. *Strategic speed bump construction*

   You use the points you have earned to buy building materials. Your team must fold and strategically place speed bumps to keep an opponent's marble from rolling into a tray where it earns points. Your Roller will try to earn points for your team by successfully rolling six marbles. Any points earned are added to your team's total. All points will be used during the Slow Marble Race culmination.

The Slow Marble Race is your final chance to have team success. Your team builds a track that allows a marble to descend "slowly" to the floor from a starting point three inches off the ground. The points you have earned buy the team the track-building materials. Your team then designs and builds a track for the final competition.

During *Fraction Speed Bumps*, you will strengthen your fraction skills and speed. All teams will need to manipulate values and compute specific answers. Pay attention to the Student Guide's helpful hints and your teacher's review lessons. Work cooperatively with your team to maximize your learning and minimize your errors.

Will you work quickly and without error? Will your team earn enough points to build your "slow marble" racetrack? Will you be more confident with fractions when you are finished? You can make the answers be "yes." Build on what you already know and learn from the success of your teammates as you attempt to navigate *Fraction Speed Bumps*.

Roles and Responsibilities

The most successful teams have team members that work together and help each other gain confidence and knowledge. (See Cooperative Group Work Rubric.)

There are various team jobs and the team members change roles each session:

Recorder

- Sits in the middle seat each day
- Fills in score sheet
- Uses calculator to convert fractions to decimals as needed

Setter

- Places symbols and values on opponent's sheet according to the instructions
- Fills in values for the team as time runs out

Checker

- Double-checks problems submitted by opponent, verifying the answer (may use calculator if needed)
- Checker may ask a teammate for help in checking a difficult problem

Roller—all team members

- Rolls marble in various situations
- Each time your team must roll marbles, a different team member must roll. Start with the Recorder, then the Setter, and finally the Checker. Once all three team members have rolled once, then the team may decide who will roll the marbles (this will cost the team 5 points).

Team Building

Teamwork helps members gain confidence and knowledge. As a team, you work on math problems, share answers, discuss strategies, and formulate plans. During construction projects, you brainstorm ideas, compromise on designs, reflect on previous successes, and even use past information to modify plans during competition. Sharing responsibility, cooperating, and focusing on doing your job each day will maximize your team's experience. Successful teams work together accurately.

Team Competition

Each team competes against the daily fraction math challenge and sometimes competes against a designated opponent for that day. Teams have different opponents on different days. Sometimes teams compete directly (head to head), but other times you only need to verify that the opponent's math is computed correctly. Finding mistakes in their work may earn you points.

How to Succeed

1. Slide an answer to the next circle. After using some of the values, you will compute an answer and place it in a circle. That answer then slides to the circle below to become part of a different math problem.

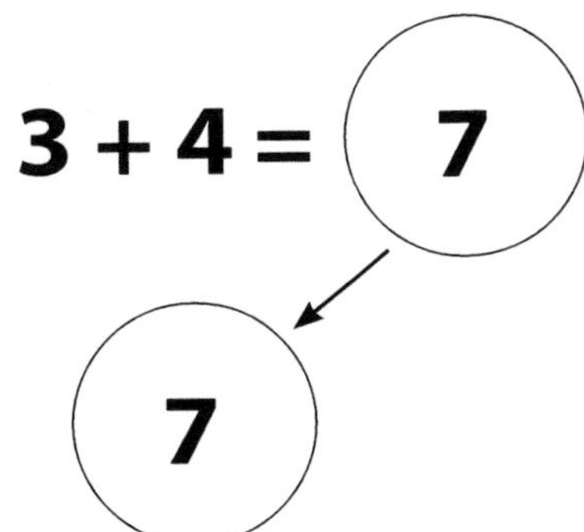

2. Place values and symbols correctly. Your team should discuss where it plans to place the values and symbols on an opponent's paper. If the team cannot agree on a strategy, the Setter for that day has the final say and must write the value or symbol for the team.
3. Select "largest" or "smallest." All players try combinations to find the best solution. Sometimes the team must discuss whether its wants to try for the largest or smallest answer. The Recorder has the job of circling the option (largest or smallest) that the team chooses and filling in the values on the team sheet for its final answer on each challenge.
4. Verify the opponent's answer. The Checker should carefully check the work on an opponent's team sheet. Remember: the Checker may ask a teammate for help in checking a problem. Finding their mistakes can benefit your team.
5. Change your Roller. The job of Roller changes each time you have a speed bump event. After three events, the team may choose who will roll the marble(s) for the next event (the cost is five points).
6. Show all work. Your teacher may award bonus points if all team members show work on their individual work papers.
7. Double-check the final answer written on the team paper by the Recorder
8. Fill in the score sheet correctly

How to Fill Out a Score Sheet

Example:

Master | Fraction Score Sheet

Fraction Score Sheet

Team Name Rocket Racers

Date	Activity	Score	Bonus	Penalty	Points used	Total
3/2	Day 2	34	7			41
3/3	Day 3	9				50
3/6	Day 4	8		−5		53
3/5	Speed bump	14			−9	58

Earning Points

There are many ways for your team to earn points:

1. Correctly working daily fraction speed bump problems
2. Finishing faster than other teams
3. Speed bump marble-rolling events
4. Bonus points:
    a. All members show work
    b. Try for the largest or smallest answer
    c. Work on a bonus problem if you finish other work quickly
5. Additional homework problems assigned by your teacher
6. Points earned when the Checker finds an opponent's mistake

Making the Track

Draw parallel lines on your heavy paper, making the space between each line 2 cm. Cut along every third line to give you three sections: the two sides and the bottom of the U-shaped track. Fold along each line and crease. You will have a rigid U-shaped length of track that can be taped to other pieces to form a continuous length of track.

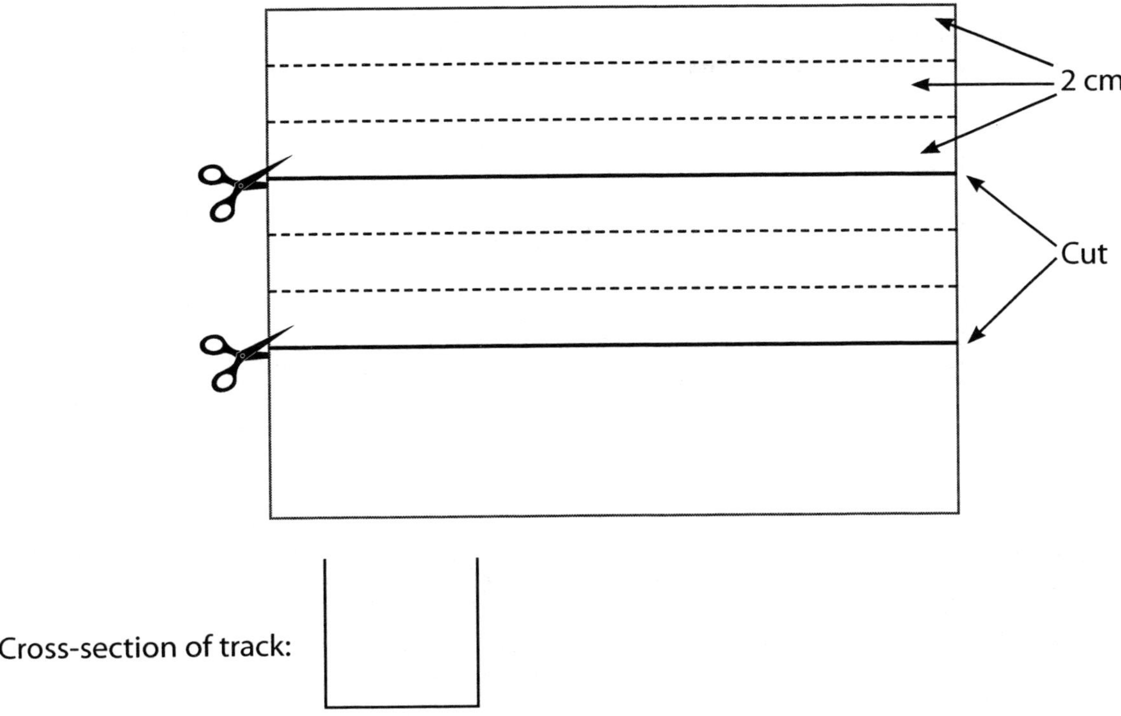

Cross-section of track:

Your track may have turns in it. The turns may be at any angle. If you want a 90-degree angle, use the directions provided below.

Cutting and folding the track to make a 90-degree turn

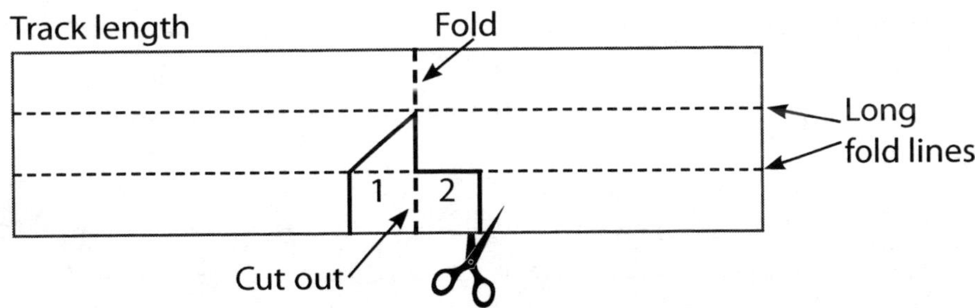

Determine where you want the 90-degree turn to be located. Draw a perpendicular line across all three sections. Draw two boxes (2 cm x 2 cm) on each side of the line. Draw a diagonal line from the first fold line to the second fold line, as shown in the diagram. Cut out the irregular shape indicated by the dark lines. Fold the track lengthwise along the regular fold lines and then fold at the blue fold line to make your right-angle turn. A right turn is shown. If you want a left turn, you must start on the other side of the track. When folding the track for the turn, have the diagonal cut end up on top of the other square shape.

Making a Pillar

You make pillars to support the track by sacrificing some of your track. Determine the height of the track at which it needs a support pillar. Measure to the top of the track at that place, and cut off a piece of your track. Cut the track as shown and fold and tape, forming a strong triangular pillar.

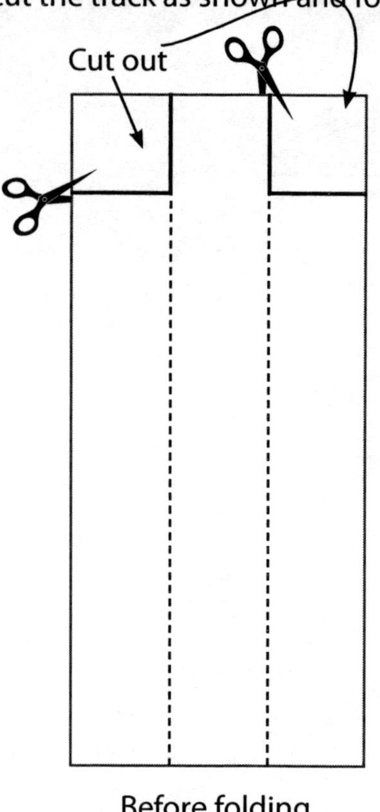

Before folding

Tab gets taped to the side of the track.

Track rests on the triangle.

Tape this edge

After folding

Definitions

- **Proper fraction:** any fraction with a numerator smaller than its denominator. It represents a value less than one.

  Examples: $\frac{1}{3}$, $\frac{1}{50}$, $\frac{7}{8}$, $\frac{98}{100}$

- **Improper fraction:** any fraction with a numerator larger than its denominator. It represents a value greater than one.

  Examples: $\frac{5}{3}$, $\frac{25}{24}$, $\frac{8}{2}$

Fraction Information

1. What is a fraction?

$\frac{3}{8}$

3 ← pieces available—numerator tells "how many"

8 ← label and the total number of pieces needed for one whole—denominator

It represents less than one whole. There are only 3 parts out of the 8 total parts.

■	■	■	

2. Every fraction is really just a division problem:

$\frac{9}{2}$—This improper fraction asks, "How many twos are in nine?"

2+2+2+2+1 = 9 The answer is $4\frac{1}{2}$.
Four groups of two and one (which is half of two)

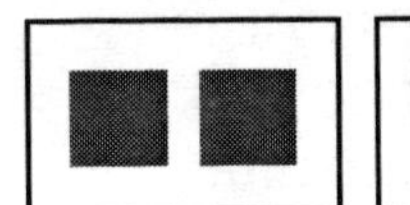  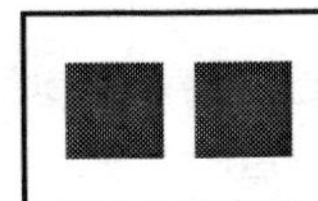 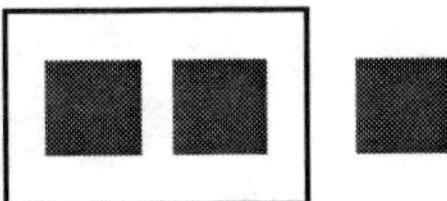

$\frac{2}{3}$—This proper fraction asks, "How many threes are in two?" There is not one whole group of three in two, so your answer will be less than one. If you divide 2 by 3 the answer is .666 (two thirds).

3. Every whole number can be changed to a fraction:

Just multiply the whole number by "one" written as a fraction ($\frac{1}{1}$, $\frac{2}{2}$, etc.)

$3 \times \frac{1}{1} = \frac{3}{1}$ ←The 1 is the number representing a "whole."

A "whole number" is written as a numerator, and we often just do not place the "1" under it, because one times anything gives you the same answer (the "One Principle").

Example: $\frac{12}{1}$ is written "12" because it is simpler to ignore the 1 in the denominator.

4. You may only add or subtract fractions with like denominators:

A denominator is just a label. You may only add and subtract when the two fractions have the same label (denominator).

$\frac{1}{3 \text{ third}} + \frac{1}{3 \text{ third}} = \frac{2}{3 \text{ thirds}}$ (same label)

$\frac{1}{\text{fish}} + \frac{1}{\text{fish}} = \frac{2}{\text{fish}}$ (same label)

7	−	2	=	5	← This is the math problem.
tenths		tenths		tenths	← This is only the label.

5. If the labels are different, when you add or subtract, you must change one or both so that you have the same labels:

$\frac{3}{4}$ (fourths) $\frac{1}{8}$ (eighths)

The labels are NOT the same, so you may not add yet.

So how can you change the labels?

The math rule is that you can multiply any number or fraction by 1 and still have the same value (answer). A fraction's label may be changed by multiplying the fraction by 1. The value 1 may also be written as a fraction.

For example: $\frac{2}{2}$ $\frac{3}{3}$ $\frac{4}{4}$ $\frac{5}{5}$, etc.

To change the fourths label to eighths, it must be multiplied by $\frac{2}{2}$, which is really 1.

$\frac{3}{4} \times \frac{2}{2} = \frac{6}{8}$ ← the label needed (This fraction, $\frac{2}{2}$, is really 1.)

$\frac{6}{8} + \frac{1}{8} = \frac{7}{8}$ ← This is the math problem. ← Labels are now the same and can be added.

6. In a mixed numeral subtraction problem, you may need to regroup:

$$\begin{array}{r} 4\frac{1}{4} \\ -\ 2\frac{3}{4} \\ \hline \end{array}$$

You may not subtract yet because you need to regroup.

In order to proceed, you must increase the top fractional amount so that you can subtract the bottom $\frac{3}{4}$. You must take one whole from the 4, making it now a 3. Then you must cash in the whole for some fractional amount. You need more fourths. One whole is equal to $\frac{4}{4}$. Just add that $\frac{4}{4}$ to the $\frac{1}{4}$ at the top to get a larger value of $\frac{5}{4}$. Now you can subtract the $\frac{3}{4}$ from the $\frac{5}{4}$.

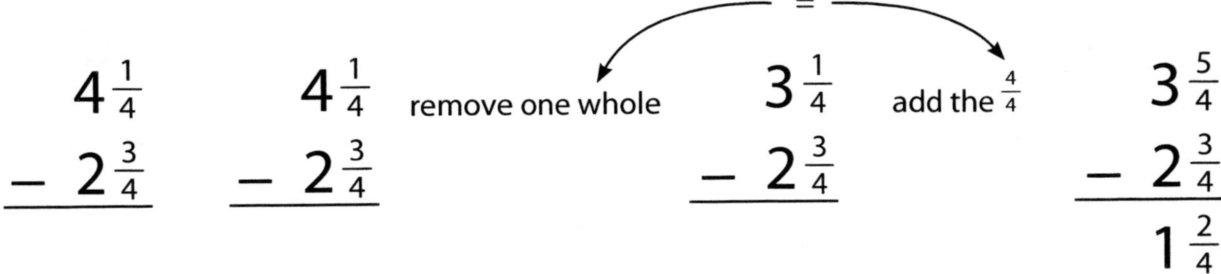

7. In a word problem, "of" means "multiply," and "is" means "equal to":

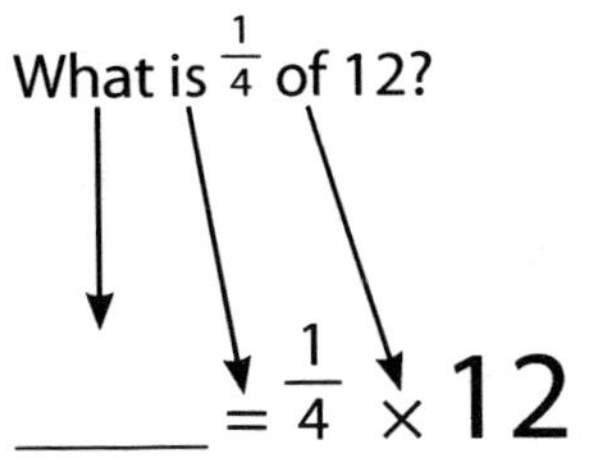

$\frac{1}{4}$ of 12 is what?

$\frac{1}{4} \times 12 =$ _____

8. When you multiply any value using a proper fraction, your answer will be a smaller part of the original value:

As long as one factor is a proper fraction you will have a smaller answer than the other factor.

Start with		multiplier		product
5	×	$\frac{1}{2}$	=	$2\frac{1}{2}$
3	×	$\frac{1}{3}$	=	1
9	×	$\frac{1}{4}$	=	$2\frac{1}{4}$
$\frac{1}{8}$	×	$\frac{1}{8}$	=	$\frac{1}{64}$
$\frac{1}{10}$	×	$\frac{1}{2}$	=	$\frac{1}{20}$

smaller than original value

The answer is always less than the original value.

9. Dividing is just asking this simple question: "How many of these are in that?"

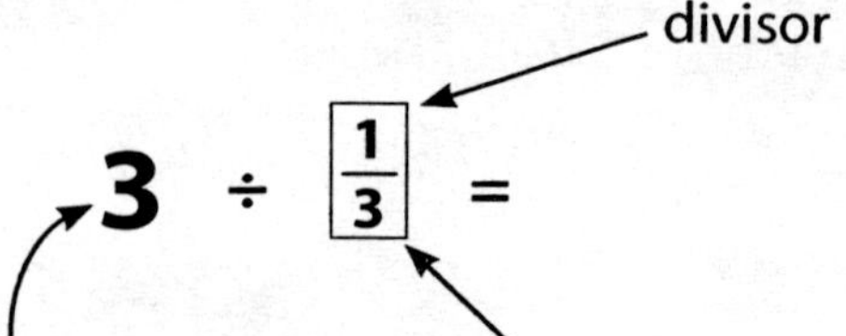

How many of these are in that?

How many one-thirds are in three?

There are nine thirds in three, because $9 \times \frac{1}{3} = 3$.

In most cases, if the divisor is smaller than the dividend (beginning value), your answer will be larger than one.

Example: $6 \div \frac{1}{2} = 12$ Since $12 \times \frac{1}{2} = 6$, there are 12 halves in 6. The answer is larger than one.

If the divisor is larger than the dividend (beginning value) your answer will be less than one.

Example: $\frac{1}{4} \div 2 = \frac{1}{8}$ Since $\frac{1}{8} \times 2 = \frac{1}{4}$, there is only one eighth of 2 in $\frac{1}{4}$. The answer is smaller than one.

In many problems, the beginning value will contain many of the other fraction parts, so the answer will be large.

Example: $10 \div \frac{1}{8} =$ So, how many $\frac{1}{8}$s are there in 10? Many.

Since $80 \times \frac{1}{8} = 10$, there are 80 eighths in 10.

10. It is easy to compare fractions that have the same numerator:

If ***the numerators are the same,*** the fraction with the largest denominator is the smaller fraction. A larger denominator means that the pieces are smaller.

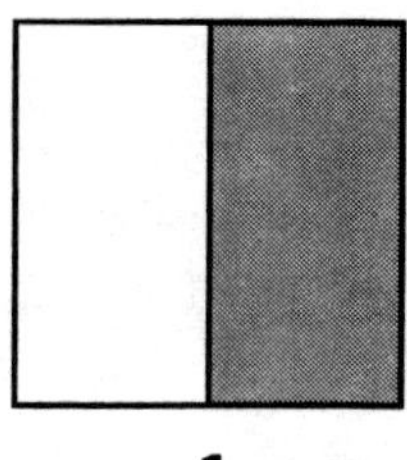

$\frac{1}{2}$

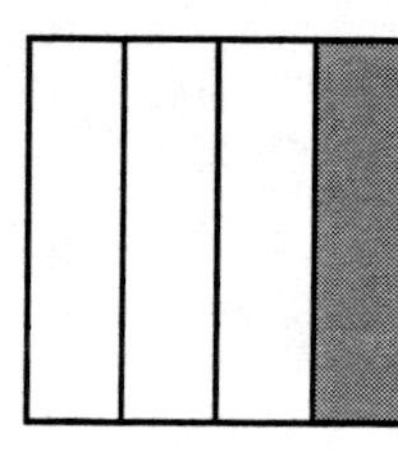

$\frac{1}{4}$

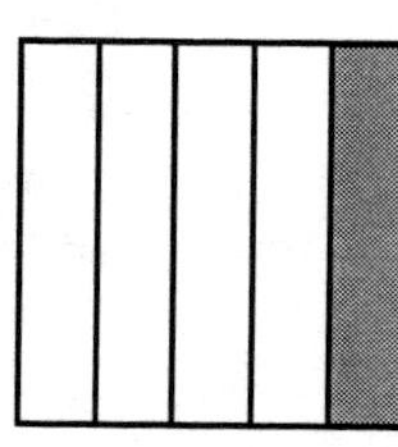

$\frac{1}{5}$

This is the smallest fraction, even though it has the largest denominator.

11. You can quickly compare many fractions just by looking at them:

The fraction with a numerator close to the value of the denominator, is usually the larger fraction.

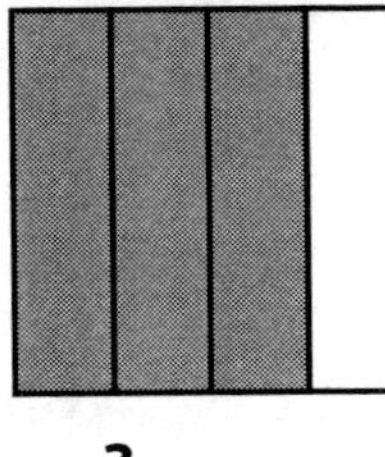

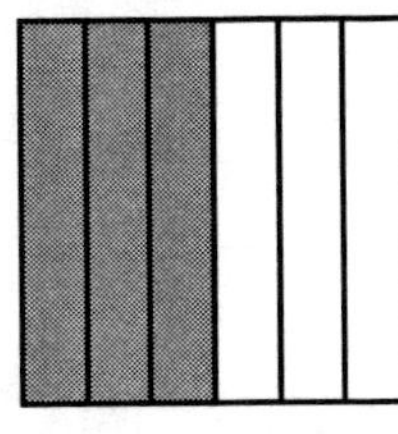

$$\frac{3}{4} > \frac{3}{6}$$

Three is almost all of (the label) four. Three is only half of (the label) six.

$$\frac{8}{11} > \frac{40}{70}$$

Eight is much more than half of 11. Forty is just a little more than half of 70.

12. Find the least common denominator of prime numbers by multiplying:

$$\frac{1}{3} + \frac{2}{5} + \frac{3}{7} =$$

prime numbers

Three, five, and seven are all prime numbers. $3 \times 5 \times 7 = 105$. That is the least common denominator for this problem.

13. Multiplication is the exact opposite of division, and division is the exact opposite of multiplication (as you can see in these fact families):

Division:

$8 \div \frac{1}{2} = 16$ "How many halves in eight?" The answer is 16.

$8 = \frac{1}{2} \times 16$

Multiplication:

$6 \times \frac{1}{3} = 2$ One third of six is two.

$6 = 2 \div \frac{1}{3}$ "How many thirds are in two?" The answer is six.

14. Reduce fractions when you can before beginning to solve a problem:

If you see a problem like $\frac{30}{60} + \frac{75}{100}$, reduce the fractions to lowest terms before adding. $\frac{30}{60}$ is really $\frac{1}{2}$, and $\frac{75}{100}$ is really $\frac{3}{4}$.

Which looks easier to work on: $\frac{30}{60} + \frac{75}{100}$ or $\frac{1}{2} + \frac{1}{4}$?

If you reduce before you begin, quite often you will not have to reduce your answer because it will be in lowest terms.

15. You can speed up your multiplying by shortcut cancelling:

Many fractions can be divided by 1 ($\frac{2}{2}$, $\frac{3}{3}$, etc.) to get a reduced fraction:

$$\frac{2}{8} \div \frac{2}{2} = \frac{1}{4}$$

In this problem ($\frac{2}{4} \times \frac{4}{8}$), you can simplify the fractions before multiplying.

Find a numerator and denominator pair that can be reduced by dividing by a fraction representing 1 ($\frac{2}{2}$, $\frac{3}{3}$, $\frac{4}{4}$, etc.).

The 4s can be divided by 4. 4 divided by 4 =1.

The 2 and the 8 can be divided by 2. 2 divided by 2 =1, and 8 divided by 2 = 4.

$$\frac{2}{4} \times \frac{4}{8} = \frac{2}{1} \times \frac{1}{8} = \frac{1}{1} \times \frac{1}{4}$$

After dividing, the problem looks like this.

When you multiply, the answer will already be reduced for you because of the shortcut cancelling.

In a longer problem, shortcut cancelling minimizes having to reduce the answer, and you can cancel numbers that are not next to each other:

$$\frac{\cancel{2}_1}{\cancel{8}_4} \times \frac{\cancel{4}_1}{\cancel{10}_2} \times \frac{\cancel{5}_1}{\cancel{8}_2} \times \frac{\cancel{6}_1}{\cancel{12}_2} = \frac{1}{32}$$ Already reduced

Study the fractions in every problem, looking for ways to easily reduce the fractions before multiplying. Notice that all of the fractions in the problem below are all equal to $\frac{1}{2}$. The numerator is exactly one half of the denominator. All of the fractions are $\frac{1}{2}$.

$$\frac{2}{4} \times \frac{4}{8} \times \frac{10}{20} \times \frac{15}{30} =$$ This difficult problem is really **easy.**

Your new easier problem looks like this:

$$\frac{1}{2} \times \frac{1}{2} \times \frac{1}{2} \times \frac{1}{2} = \frac{1}{16}$$

16. Compare fractions by dividing the numerator by the denominator:

Which is larger: $\frac{3}{14}$ or $\frac{6}{26}$?

Divide 3 by 14...you get the decimal 0.214.
Divide 6 by 26...you get the decimal 0.231.

Therefore, $\frac{6}{26}$ is the larger fraction.

17. Least common multiple: Change fractions to the same label by finding their least common multiple. The least common multiple is the smallest number that both denominators will divide into evenly.

$$\frac{2}{4} + \frac{1}{8} =$$

$\frac{2}{4}$ and $\frac{1}{8}$ can not be added until they have the same labels. Fourths and eighths are not the same.

First, check to see if the smaller denominator (4) will divide evenly into the larger denominator (8). Since 4 can divide evenly into 8, the new label for both will be "eighths."

Multiply the $\frac{2}{4}$ by $\frac{2}{2}$ (one) to get an equivalent fraction with the label of "eighths."

$$\frac{1}{3} + \frac{3}{4} =$$

If one of the two denominators is a prime number and that prime number will not divide evenly into the other denominator, then multiply the two denominators to get the least common multiple. That will be your new label for both fractions.

$3 \times 4 = 12$. Your new label is "twelfths."

Change both fractions to equivalent fractions with "twelfths" as their labels. Add.

$$\frac{2}{7} + \frac{3}{5} =$$

If both fractions are prime numbers, just multiply them to find the least common multiple, which will be the new label.

$7 \times 5 = 35$. Your new label is "thirty-fifths."

Change both fractions to equivalent fractions with "thirty-fifths" as their labels. Add.

$$\frac{3}{4} + \frac{2}{5} =$$

Sometimes, you can take the larger denominator and double it, or triple it, or quadruple it until you find a value that the small denominator will divide into evenly. That will be your new label for the problem.

Double 5 = 10

Triple 5 = 15

Quadruple 5 = 20. This is the first number that four will divide into evenly. It is the new label for this problem.

Change both fractions to equivalent fractions with "twentieths" as their labels. Add.

18. **Greatest common factor:** The greatest common factor is the largest number that will divide into both the numerator and denominator of a fraction. Use the greatest common factor when reducing fractions to lowest terms.

In $\frac{4}{12}$, 4 is the largest number that will divide into both 4 and 12 evenly. 4 is the greatest common factor you can use to reduce this fraction.

In $\frac{6}{15}$, 3 is the largest number that will divide into both 6 and 15 evenly. 3 is the greatest common factor you can use to reduce this fraction.

When you can not figure out the greatest common factor by looking at the problem, you can find it by writing the prime factors of both the numerator and the denominator:

$\frac{12}{30}$

The prime factors of 12 are $2 \times 2 \times 3$.

The prime factors of 30 are $2 \times 3 \times 5$.

The two numbers (12 and 30) both share the factors 2 and 3. The product of 2×3 is 6, the greatest common factor for 12 and 30, and is used to reduce the fraction to lowest terms by dividing:

$12 \div 6 = 2$

$30 \div 6 = 5$ $\frac{12}{30}$ reduces to $\frac{2}{5}$

Teacher Feedback Form

At Interact, we constantly strive to make our units the best they can be. We always appreciate feedback from you—our customer—to facilitate this process. With your input, we can continue to provide high-quality, interactive, and meaningful instructional materials to enhance your curriculum and engage your students. Please take a few moments to complete this feedback form and drop it in the mail. Address it to:

Interact • Attn: Editorial
10200 Jefferson Blvd. • P.O. Box 802
Culver City, CA 90232-0802

or fax it to us at **(800) 944-5432**

or e-mail it to us at **access@teachinteract.com**

We enjoy receiving photos or videos of our units in action!
Please use the release form on the following page.

Your name: ______________________________

Address: ______________________________

____

E-mail: ______________________________

Interact unit: ______________________________

Comments: ______________________________

____

____

____

____

____

____

____

____

Release Form for Photographic Images

To Teachers:

To help illustrate to others the experiential activities involved and to promote the use of simulations, we like to get photographs and videos of classes participating in the simulation. Please send photos of students actively engaged so we can publish them in our promotional material. Be aware that we can only use images of students for whom a release form has been submitted.

To Parents:

I give permission for photographs or videos of my child to appear in catalogs of educational materials published by Interact.

Name of student: ______________________________ (print)

Age of student: ______________________________ (print)

Parent or guardian: ______________________________ (print)

Signature: ______________________________ Date: ______________

Address:

____

____

____

____

Phone: ______________________________

Interact
10200 Jefferson Blvd.
Culver City, CA 90232-0802
310-839-2436